Let's Keep in Touch

Follow Us   Online

Visit US at

www.EffortlessMath.com

                                    https://www.facebook.com/Effortlessmath

Call

   https://goo.gl/2B6qWW

1-469-230-3605

Online Math Lessons

It's easy! Here's how it works.

1- Request a FREE introductory session.

2- Meet a Math tutor online.

3- Start Learning Math in Minutes.

Send Email to: info@EffortlessMath.com

Or Call: **+1-469-230-3605**

www.EffortlessMath.com

... So Much More Online!

- **FREE Math lessons**

- **More Math learning books!**

- **Online Math Tutors**

Looking for an Online Math Tutor?

Call us at: 001-469-230-3605

Send email to: Info@EffortlessMath.com

7th Grade PARCC Math Workbook 2018:

The Most Comprehensive Review for the Math Section of the PARCC TEST

By

Reza Nazari

& Ava Ross

Copyright © 2018

Reza Nazari & Ava Ross

All rights reserved. No part of this publication may be reproduced, stored in a retrieval system, or transmitted in any form or by any means, electronic, mechanical, photocopying, recording, scanning, or otherwise, except as permitted under Section 107 or 108 of the 1976 United States Copyright Ac, without permission of the author.

All inquiries should be addressed to:

info@effortlessMath.com

www.EffortlessMath.com

ISBN-13: 978-1985411715

ISBN-10: 1985411717

Published by: Effortless Math Education

www.EffortlessMath.com

7th Grade PARCC Math Workbook 2018

Description

Effortless Math PARCC Workbook provides students with the confidence and math skills they need to succeed on the PARCC Math, providing a solid foundation of basic Math topics with abundant exercises for each topic. It is designed to address the needs of PARCC test takers who must have a working knowledge of basic Math.

This comprehensive workbook with over 2,500 sample questions and 2 complete 7th Grade PARCC tests is all your student needs to fully prepare for the PARCC Math. It will help your student learns everything they need to ace the math section of the PARCC.

There are more than 2,500 Math problems with answers in this book.
Effortless Math unique study program provides your student with an in-depth focus on the math portion of the exam, helping them master the math skills that students find the most troublesome.
This workbook contains most common sample questions that are most likely to appear in the mathematics section of the PARCC.

Inside the pages of this comprehensive workbook, students can learn basic math operations in a structured manner with a complete study program to help them understand essential math skills. It also has many exciting features, including:

- Dynamic design and easy-to-follow activities
- A fun, interactive and concrete learning process
- Targeted, skill-building practices
- Fun exercises that build confidence
- Math topics are grouped by category, so the students can focus on the topics they struggle on
- All solutions for the exercises are included, so you will always find the answers
- 2 Complete PARCC Math Practice Tests that reflect the format and question types on PARCC

Effortless Math PARCC Workbook is an incredibly useful tool for those who want to review all topics being covered on the PARCC test. It efficiently and effectively reinforces learning outcomes through engaging questions and repeated practice, helping students to quickly master basic Math skills.

About the Author

Reza Nazari is the author of more than 50 Math learning books including:
- **Math for Super Smart Students:** Fifth Graders and Older by Reza Nazari
- **Math and Critical Thinking Challenges:** For the Middle and High School Student
- **Effortless Math Education Workbooks**

Reza is also an experienced Math instructor and a test-prep expert who has been tutoring students since 2008. Reza is the founder of Effortless Math Education, a tutoring company that has helped many students raise their standardized test scores--and attend the colleges of their dreams. Reza provides an individualized custom learning plan and the personalized attention that makes a difference in how students view math.

To ask questions about Math, you can contact Reza via email at:
reza@EffortlessMath.com

Find Reza's professional profile at:
goo.gl/zoC9rJ

Contents

Description 2

Chapter 1: Arithmetic and Number Theory 10

1–1 Simplifying Fractions 11

1–2 Adding and Subtracting Fractions 12

1–3 Multiplying and Dividing Fractions 13

1–4 Adding and Subtracting Mixed Numbers 14

1–5 Multiplying and Dividing Mixed Numbers 15

1–6 Adding and Subtracting Decimals 16

1–7 Multiplying and Dividing Decimals 17

1–8 Converting Between Fractions, Decimals and Mixed Numbers 18

1–9 Rounding Numbers 19

1–10 Factoring Numbers 20

1–11 Prime Factorization 21

1–12 Greatest Common Factor 22

1–13 Least Common Multiple 23

1–14 Divisibility Rules 24

Answers of Worksheets – Chapter 1 25

Chapter 2: Real Numbers and Integers 32

2–1 Adding and Subtracting Integers 33

2–2 Multiplying and Dividing Integers 34

2–3 Ordering Integers and Numbers 35

2–4 Arrange, Order, and Comparing Integers 36

2–5 Order of Operations 37

2–6 Mixed Integer Computations 38

2–7 Absolute Value 39

2–8 Integers and Absolute Value 40

2–9 Classifying Real Numbers Venn Diagram 41

Answers of Worksheets – Chapter 2 42

Chapter 3: Proportions and Ratios 45

3–1 Writing Ratios 46

3–2 Simplifying Ratios 47

3–3 Proportional Ratios 48

3–4 Create a Proportion 49

3–5 Similar Figures 50

3–6 Similar Figure Word Problems 51

3–7 Complete the Ratio Table 52

3–8 Ratio and Rates Word Problems 53

Answers of Worksheets – Chapter 3 54

Chapter 4: Percent 57

4–1 Converting Between Percent, Fractions, and Decimals 58

4–2 Table of Common Percent 59

4–3 Percentage Calculations 60

4–4 Find What Percentage a Number Is of Another 61

4–5 Find a Percentage of a Given Number 62

4–6 Percent Problems 63

4–7 Percent of Increase and Decrease 64

4–8 Markup, Discount, and Tax 65

Answers of Worksheets – Chapter 4 66

Chapter 5: Algebraic Expressions 70

5–1 Expressions and Variables 71

5–2 Simplifying Variable Expressions 72

5–3 The Distributive Property 73

5–4 Translate Phrases into an Algebraic Statement 74

5–5 Evaluating One Variable 75

5–6 Evaluating Two Variables 76

5–7 Combining like Terms 77

5–8 Simplifying Polynomial Expressions 78

Answers of Worksheets – Chapter 5 79

Chapter 6: Equations 82

6–1 One–Step Equations 83

6–2 One–Step Equation Word Problems 84

6–3 Two–Step Equations 85

6–4 Two–Step Equation Word Problems 86

6–5 Multi–Step Equations 87

Answers of Worksheets – Chapter 6 88

Chapter 7: Systems of Equations 90

7–1 Solving Systems of Equations by Graphing 91

7–2 Solving Systems of Equations by Substitution 92

7–3 Solving Systems of Equations by Elimination 93

7–4 Systems of Equations Word Problems 94

Answers of Worksheets – Chapter 7 95

Chapter 8: Inequalities 97

8–1 Graphing Single – Variable Inequalities 98

8–2 One–Step Inequalities 99

8–3 Two–Step Inequalities 100

8–4 Multi–Step Inequalities 101

Answers of Worksheets – Chapter 8 102

Chapter 9: Linear Functions 105

9–1 Finding Slope 106

9–2 Graphing Lines Using Slope–Intercept Form 107

9–3 Graphing Lines Using Standard Form 108

9–4 Writing Linear Equations 109

9–5 Graphing Linear Inequalities 110

Answers of Worksheets – Chapter 9 111

Chapter 10: Exponents and Radicals 116

10–1 Multiplication Property of Exponents 117

10–2 Division Property of Exponents 118

10–3 Powers of Products and Quotients 119

10–4 Zero and Negative Exponents 120

10–5 Negative Exponents and Negative Bases 121

10–6 Writing Scientific Notation 122

10–7 Square Roots 123

Answers of Worksheets – Chapter 10 124

Chapter 11: Plane Figures 127

11–1 Transformations: Translations, Rotations, and Reflections 128

11–2 The Pythagorean Theorem 129

11–3 Classifying Triangles and Quadrilaterals 130

11–4 Area of Triangles 131

11–5 Perimeter of Polygons 132

11–6 Area and Circumference of Circles 133

11–7 Area of Squares, Rectangles, and Parallelograms 134

11–8 Area of Trapezoids 135

Answers of Worksheets – Chapter 11 136

Chapter 12: Solid Figures 139

12–1 Classifying Solids 140

12–2 Volume of Cubes and Rectangle Prisms 141

12–3 Surface Area of Cubes 142

12–4 Surface Area of a Prism 143

Answers of Worksheets – Chapter 12 144

Chapter 13: Statistics 145

13–1 Mean, Median, Mode, and Range of the Given Data 146

13–2 First Quartile, Second Quartile and Third Quartile of the Given Data 147

13–3 Bar Graph 148

13–4 Box and Whisker Plots 149

13–5 Stem–And–Leaf Plot 150

13–6 The Pie Graph or Circle Graph 151

13–7 Scatter Plots 152

Answers of Worksheets – Chapter 13 153

All Mathematics Formulas a 7th grade student Must Know! 157

Mathematics Formula Sheet 158

PARCC Practice Test 1 176

PARCC Practice Test 2 195

PARCC Practice Tests Answer keys 220

Chapter 1: Arithmetic and Number Theory

1–1 Simplifying Fractions

1–2 Adding and Subtracting Fractions

1–3 Multiplying and Dividing Fractions

1–4 Adding and Subtracting Mixed Numbers

1–5 Multiplying and Dividing Mixed Numbers

1–6 Adding and Subtracting Decimals

1–7 Multiplying and Dividing Decimals

1–8 Converting Between Fractions, Decimals and Mixed Numbers

1–9 Rounding Numbers

1–10 Factoring Numbers

1–11 Prime Factorization

1–12 Greatest Common Factor

1–13 Least Common Multiple

1–14 Divisibility Rules

1–1 Simplifying Fractions

Simplify the fractions.

1) $\dfrac{22}{36}$

2) $\dfrac{8}{10}$

3) $\dfrac{12}{18}$

4) $\dfrac{6}{8}$

5) $\dfrac{13}{39}$

6) $\dfrac{5}{20}$

7) $\dfrac{16}{36}$

8) $\dfrac{18}{36}$

9) $\dfrac{20}{50}$

10) $\dfrac{6}{54}$

11) $\dfrac{45}{81}$

12) $\dfrac{21}{28}$

13) $\dfrac{35}{56}$

14) $\dfrac{52}{64}$

15) $\dfrac{13}{65}$

16) $\dfrac{44}{77}$

17) $\dfrac{21}{42}$

18) $\dfrac{15}{36}$

19) $\dfrac{9}{24}$

20) $\dfrac{20}{80}$

21) $\dfrac{25}{45}$

1–2 Adding and Subtracting Fractions

Add fractions.

1) $\dfrac{2}{3} + \dfrac{1}{2}$

2) $\dfrac{3}{5} + \dfrac{1}{3}$

3) $\dfrac{5}{6} + \dfrac{1}{2}$

4) $\dfrac{7}{4} + \dfrac{5}{9}$

5) $\dfrac{2}{5} + \dfrac{1}{5}$

6) $\dfrac{3}{7} + \dfrac{1}{2}$

7) $\dfrac{3}{4} + \dfrac{2}{5}$

8) $\dfrac{2}{3} + \dfrac{1}{5}$

9) $\dfrac{16}{25} + \dfrac{3}{5}$

Subtract fractions.

10) $\dfrac{4}{5} - \dfrac{2}{5}$

11) $\dfrac{3}{5} - \dfrac{2}{7}$

12) $\dfrac{1}{2} - \dfrac{1}{3}$

13) $\dfrac{8}{9} - \dfrac{3}{5}$

14) $\dfrac{3}{7} - \dfrac{3}{14}$

15) $\dfrac{4}{15} - \dfrac{1}{10}$

16) $\dfrac{3}{4} - \dfrac{13}{18}$

17) $\dfrac{5}{8} - \dfrac{2}{5}$

18) $\dfrac{1}{2} - \dfrac{1}{9}$

1–3 Multiplying and Dividing Fractions

Multiplying fractions. Then simplify.

1) $\dfrac{1}{5} \times \dfrac{2}{3}$

2) $\dfrac{3}{4} \times \dfrac{2}{3}$

3) $\dfrac{2}{5} \times \dfrac{3}{7}$

4) $\dfrac{3}{8} \times \dfrac{1}{3}$

5) $\dfrac{3}{5} \times \dfrac{2}{5}$

6) $\dfrac{7}{9} \times \dfrac{1}{3}$

7) $\dfrac{2}{3} \times \dfrac{3}{8}$

8) $\dfrac{1}{4} \times \dfrac{1}{3}$

9) $\dfrac{5}{7} \times \dfrac{7}{12}$

Dividing fractions.

10) $\dfrac{2}{9} \div \dfrac{1}{4}$

11) $\dfrac{1}{2} \div \dfrac{1}{3}$

12) $\dfrac{6}{11} \div \dfrac{3}{4}$

13) $\dfrac{11}{14} \div \dfrac{1}{10}$

14) $\dfrac{3}{5} \div \dfrac{5}{9}$

15) $\dfrac{1}{2} \div \dfrac{1}{2}$

16) $\dfrac{3}{5} \div \dfrac{1}{5}$

17) $\dfrac{12}{21} \div \dfrac{3}{7}$

18) $\dfrac{5}{14} \div \dfrac{9}{10}$

1–4 Adding and Subtracting Mixed Numbers

Add.

1) $4\frac{1}{2} + 5\frac{1}{2}$

2) $2\frac{3}{8} + 3\frac{1}{8}$

3) $5\frac{3}{5} + 5\frac{1}{5}$

4) $1\frac{1}{3} + 2\frac{2}{3}$

5) $5\frac{1}{6} + 5\frac{1}{2}$

6) $3\frac{1}{3} + 1\frac{1}{3}$

7) $1\frac{10}{11} + 1\frac{1}{3}$

8) $2\frac{3}{6} + 1\frac{1}{2}$

9) $5\frac{3}{5} + 5\frac{1}{5}$

10) $7 + \frac{1}{5}$

11) $1\frac{5}{7} + \frac{1}{3}$

12) $2\frac{1}{4} + 1\frac{2}{4}$

Subtract.

13) $4\frac{1}{2} - 3\frac{1}{2}$

14) $3\frac{3}{8} - 3\frac{1}{8}$

15) $6\frac{3}{5} - 5\frac{1}{5}$

16) $2\frac{1}{3} - 1\frac{2}{3}$

17) $6\frac{1}{6} - 5\frac{1}{2}$

18) $3\frac{1}{3} - 1\frac{1}{3}$

19) $2\frac{10}{11} - 1\frac{1}{3}$

20) $2\frac{1}{2} - 1\frac{1}{2}$

21) $6\frac{3}{5} - 2\frac{1}{5}$

22) $7\frac{2}{5} - 1\frac{1}{5}$

23) $2\frac{5}{7} - 1\frac{1}{3}$

24) $2\frac{1}{4} - 1\frac{1}{2}$

1–5 Multiplying and Dividing Mixed Numbers

Find each product.

1) $1\frac{2}{3} \times 1\frac{1}{4}$

2) $1\frac{3}{5} \times 1\frac{2}{3}$

3) $1\frac{2}{3} \times 3\frac{2}{7}$

4) $4\frac{1}{8} \times 1\frac{2}{5}$

5) $2\frac{2}{5} \times 3\frac{1}{5}$

6) $1\frac{1}{3} \times 1\frac{2}{3}$

7) $1\frac{5}{8} \times 2\frac{1}{2}$

8) $3\frac{2}{5} \times 2\frac{1}{5}$

9) $2\frac{2}{3} \times 4\frac{1}{4}$

10) $2\frac{3}{5} \times 1\frac{2}{4}$

11) $1\frac{1}{3} \times 1\frac{1}{4}$

12) $3\frac{2}{5} \times 1\frac{1}{5}$

Find each quotient.

13) $2\frac{1}{5} \div 2\frac{1}{2}$

14) $2\frac{3}{5} \div 1\frac{1}{3}$

15) $3\frac{1}{6} \div 4\frac{2}{3}$

16) $1\frac{2}{3} \div 3\frac{1}{3}$

17) $4\frac{1}{8} \div 2\frac{2}{4}$

18) $3\frac{1}{2} \div 2\frac{3}{5}$

19) $3\frac{5}{9} \div 1\frac{2}{5}$

20) $2\frac{2}{7} \div 1\frac{1}{2}$

21) $3\frac{1}{5} \div 1\frac{1}{2}$

22) $4\frac{3}{5} \div 2\frac{1}{3}$

23) $6\frac{1}{6} \div 1\frac{2}{3}$

24) $2\frac{2}{3} \div 1\frac{1}{3}$

1-6 Adding and Subtracting Decimals

Add and subtract decimals.

1) 15.14 − 12.18 = _____

2) 65.72 + 43.67 = _____

3) 82.56 + 12.28 = _____

4) 34.18 − 23.45 = _____

5) 90.37 + 56.97 = _____

6) 45.78 − 23.39 = _____

Solve.

7) _____ + 1.3 = 4.8

8) 4.2 + _____ = 11.6

9) 9.9 + _____ = 16

10) 6.9 + _____ = 16.4

11) _____ + 5.1 = 8.6

12) _____ + 7.9 = 15.2

1–7 Multiplying and Dividing Decimals

Find each product.

1) 4.5 × 1.6

2) 7.7 × 9.9

3) 2.6 × 1.5

4) 8.9 × 9.7

5) 15.1 × 12.6

6) 6.9 × 3.3

7) 5.7 × 7.8

8) 98.20 × 100

9) 23.99 × 1000

Find each quotient.

10) 9.2 ÷ 3.6

11) 27.6 ÷ 3.8

12) 12.6 ÷ 4.7

13) 6.5 ÷ 8.1

14) 1.4 ÷ 10

15) 3.6 ÷ 100

16) 4.24 ÷ 10

17) 14.6 ÷ 100

1–8 Converting Between Fractions, Decimals and Mixed Numbers

Convert fractions to decimals.

1) $\dfrac{9}{10}$

2) $\dfrac{56}{100}$

3) $\dfrac{3}{4}$

4) $\dfrac{2}{5}$

5) $\dfrac{1}{3}$

6) $\dfrac{4}{5}$

7) $\dfrac{5}{6}$

8) $\dfrac{5}{8}$

9) $\dfrac{13}{21}$

Convert decimal into fraction.

10) 0.3

11) 4.5

12) 2.5

13) 2.3

14) 0.8

15) 0.25

16) 0.14

17) 0.2

18) 0.08

19) 0.45

20) 2.6

21) 5.2

1–9 Rounding Numbers

Round each number to the underlined place value.

1) <u>9</u>72

2) 2,<u>9</u>95

3) 3<u>6</u>4

4) <u>8</u>1

5) <u>5</u>5

6) 3<u>3</u>4

7) 1,<u>2</u>03

8) 9.<u>5</u>7

9) 7.<u>4</u>84

10) 9.<u>1</u>4

11) <u>3</u>9

12) <u>9</u>,123

13) 3,4<u>5</u>2

14) <u>5</u>69

15) 1,<u>2</u>30

16) <u>9</u>8

17) <u>9</u>3

18) <u>3</u>7

19) 4<u>9</u>3

20) 2,<u>9</u>23

21) <u>9</u>,845

22) 5<u>5</u>5

23) <u>9</u>39

24) <u>6</u>9

1–10 Factoring Numbers

List all positive factors of each number.

1) 68
2) 56
3) 24
4) 40
5) 86
6) 78
7) 50
8) 98
9) 45
10) 26
11) 54
12) 28
13) 55
14) 85
15) 50

List the prime factorization for each number.

16) 50
17) 25
18) 69
19) 21
20) 45
21) 68
22) 26
23) 86
24) 93

1–11 Prime Factorization

Factor the following numbers to their prime factors.

1) 20

2) 12

3) 16

4) 27

5) 36

6) 42

7) 58

8) 35

9) 62

10) 49

11) 51

12) 78

13) 63

14) 77

15) 46

16) 69

17) 18

18) 32

19) 15

20) 33

21) 40

1–12 Greatest Common Factor

Find the GCF for each number pair.

1) 20, 30

2) 4, 14

3) 5, 45

4) 68, 12

5) 5, 12

6) 15, 27

7) 3, 24

8) 34, 6

9) 4, 10

10) 5, 3

11) 6, 16

12) 30, 3

13) 24, 28

14) 70, 10

15) 45, 8

16) 90, 35

17) 78, 34

18) 55, 75

19) 60, 72

20) 100, 78

21) 30, 40

1–13 Least Common Multiple

Find the LCM for each number pair.

1) 4, 14

2) 5, 15

3) 16, 10

4) 4, 34

5) 8, 3

6) 12, 24

7) 9, 18

8) 5, 6

9) 8, 19

10) 9, 21

11) 19, 29

12) 7, 6

13) 25, 6

14) 4, 8

15) 30, 10, 50

16) 18, 36, 27

17) 12, 8, 18

18) 8, 18, 4

19) 26, 20, 30

20) 10, 4, 24

21) 15, 30, 45

1–14 Divisibility Rules

Use the divisibility rules to find the factors of the numbers.

8	<u>2</u> 3 <u>4</u> 5 6 7 <u>8</u> 9 10
1) 16	2 3 4 5 6 7 8 9 10
2) 10	2 3 4 5 6 7 8 9 10
3) 15	2 3 4 5 6 7 8 9 10
4) 28	2 3 4 5 6 7 8 9 10
5) 36	2 3 4 5 6 7 8 9 10
6) 15	2 3 4 5 6 7 8 9 10
7) 27	2 3 4 5 6 7 8 9 10
8) 70	2 3 4 5 6 7 8 9 10
9) 57	2 3 4 5 6 7 8 9 10
10) 102	2 3 4 5 6 7 8 9 10
11) 144	2 3 4 5 6 7 8 9 10
12) 75	2 3 4 5 6 7 8 9 10

Answers of Worksheets – Chapter 1

1–1 Simplifying Fractions

1) $\frac{11}{18}$
2) $\frac{4}{5}$
3) $\frac{2}{3}$
4) $\frac{3}{4}$
5) $\frac{1}{3}$
6) $\frac{1}{4}$
7) $\frac{4}{9}$
8) $\frac{1}{2}$
9) $\frac{2}{5}$
10) $\frac{1}{9}$
11) $\frac{5}{9}$
12) $\frac{3}{4}$
13) $\frac{5}{8}$
14) $\frac{13}{16}$
15) $\frac{1}{5}$
16) $\frac{4}{7}$
17) $\frac{1}{2}$
18) $\frac{5}{12}$
19) $\frac{3}{8}$
20) $\frac{1}{4}$
21) $\frac{5}{9}$

1–2 Adding and Subtracting Fractions

1) $\frac{7}{6}$
2) $\frac{14}{15}$
3) $\frac{4}{3}$
4) $\frac{83}{36}$
5) $\frac{3}{5}$
6) $\frac{13}{14}$
7) $\frac{23}{20}$
8) $\frac{13}{15}$
9) $\frac{31}{25}$
10) $\frac{2}{5}$
11) $\frac{11}{35}$
12) $\frac{1}{6}$
13) $\frac{13}{45}$
14) $\frac{3}{14}$
15) $\frac{1}{6}$
16) $\frac{1}{36}$
17) $\frac{9}{40}$
18) $\frac{7}{18}$

1–3 Multiplying and Dividing Fractions

1) $\frac{2}{15}$

2) $\frac{1}{2}$

3) $\frac{6}{35}$

4) $\frac{1}{8}$

5) $\frac{6}{25}$

6) $\frac{7}{27}$

7) $\frac{1}{4}$

8) $\frac{1}{12}$

9) $\frac{5}{12}$

10) $\frac{8}{9}$

11) $\frac{3}{2}$

12) $\frac{8}{11}$

13) $\frac{55}{7}$

14) $\frac{27}{25}$

15) 1

16) 3

17) $\frac{4}{3}$

18) $\frac{25}{63}$

1–4 Adding and Subtracting Mixed Numbers

1) 10

2) $5\frac{1}{2}$

3) $10\frac{4}{5}$

4) 4

5) $10\frac{2}{3}$

6) $4\frac{2}{3}$

7) $3\frac{8}{33}$

8) 4

9) $10\frac{4}{5}$

10) $7\frac{1}{5}$

11) $2\frac{1}{21}$

12) $3\frac{3}{4}$

13) 1

14) $\frac{1}{4}$

15) $1\frac{2}{5}$

16) $\frac{2}{3}$

17) $\frac{2}{3}$

18) 2

19) $1\frac{19}{33}$

20) 1

21) $4\frac{2}{5}$

22) $6\frac{1}{5}$

23) $1\frac{8}{21}$

24) $\frac{3}{4}$

1–5 Multiplying and Dividing Mixed Numbers

1) $2\frac{1}{12}$
2) $2\frac{2}{3}$
3) $5\frac{10}{21}$
4) $5\frac{31}{40}$
5) $7\frac{17}{25}$
6) $2\frac{2}{9}$
7) $4\frac{1}{16}$
8) $7\frac{12}{25}$
9) $11\frac{1}{3}$
10) $3\frac{9}{10}$
11) $1\frac{2}{3}$
12) $4\frac{2}{25}$
13) $\frac{22}{25}$
14) $1\frac{19}{20}$
15) $\frac{19}{28}$
16) $\frac{1}{2}$
17) $1\frac{13}{20}$
18) $1\frac{9}{26}$
19) $2\frac{34}{63}$
20) $1\frac{11}{20}$
21) $2\frac{2}{15}$
22) $1\frac{34}{35}$
23) $3\frac{7}{10}$
24) 2

1–6 Adding and Subtracting Decimals

1) 2.96
2) 109.39
3) 94.84
4) 10.73
5) 147.34
6) 22.39
7) 3.5
8) 7.4
9) 6.1
10) 9.5
11) 3.5
12) 7.3

1–7 Multiplying and Dividing Decimals

1) 7.2
2) 76.23
3) 3.9
4) 86.33
5) 190.26
6) 22.77
7) 44.46
8) 9820
9) 23990
10) 2.5555…
11) 7.2631…
12) 2.6808…
13) 0.8024…
14) 0.14
15) 0.036
16) 0.424
17) 0.146

1–8 Converting Between Fractions, Decimals and Mixed Numbers

1) 0.9
2) 0.56
3) 0.75
4) 0.4
5) 0.333…
6) 0.8
7) 0.8333…
8) 0.625
9) 0.6190…
10) $\frac{3}{10}$
11) $4\frac{1}{2}$
12) $2\frac{1}{2}$
13) $2\frac{3}{10}$
14) $\frac{4}{5}$
15) $\frac{1}{4}$
16) $\frac{7}{50}$
17) $\frac{1}{5}$
18) $\frac{2}{25}$
19) $\frac{9}{20}$
20) $2\frac{3}{5}$
21) $5\frac{1}{5}$

1–9 Rounding Numbers

1) 1,000
2) 3,000
3) 360
4) 80
5) 60
6) 330
7) 1,200
8) 9.6
9) 7.5
10) 9.1
11) 40
12) 9,000
13) 3,450
14) 600
15) 1,200
16) 100
17) 90
18) 40
19) 490
20) 2,900
21) 10,000
22) 560
23) 900
24) 70

1–10 Factoring Numbers

1) 1, 2, 4, 17, 34, 68
2) 1, 2, 4, 7, 8, 14, 28, 56
3) 1, 2, 3, 4, 6, 8, 12, 24
4) 1, 2, 4, 5, 8, 10, 20, 40
5) 1, 2, 43, 86
6) 1, 2, 3, 6, 13, 26, 39, 78
7) 1, 2, 5, 10, 25, 50
8) 1, 2, 7, 14, 49, 98
9) 1, 3, 5, 9, 15, 45
10) 1, 2, 13, 26
11) 1, 2, 3, 6, 9, 18, 27, 54
12) 1, 2, 4, 7, 14, 28

13) 1, 5, 11, 55
14) 1, 5, 17, 85
15) 1, 2, 5, 10, 25, 50
16) 2 × 5 × 5
17) 5 × 5
18) 3 × 23
19) 3 × 7
20) 3 × 3 × 5
21) 2 × 2 × 17
22) 2 × 13
23) 2 × 43
24) 3 × 31

1–11 Prime Factorization

1) 2 . 2 . 5
2) 2 . 2 . 3
3) 2 . 2 . 2 . 2
4) 3 . 3 . 3
5) 2 . 2 . 3 . 3
6) 2 . 3 . 7
7) 2 . 29
8) 5 . 7
9) 2 . 31
10) 7 . 7
11) 3 . 17
12) 2 . 3 . 13
13) 3 . 3 . 7
14) 7 . 11
15) 2 . 23
16) 3 . 23
17) 2 . 3 . 3
18) 2 . 2 . 2 . 2 . 2
19) 3 . 5
20) 3 . 11
21) 2 . 2 . 2 . 5

1–12 Greatest Common Factor

1) 10
2) 2
3) 5
4) 4
5) 1
6) 3
7) 3
8) 2
9) 2
10) 1
11) 2
12) 3
13) 4
14) 10
15) 1
16) 5
17) 2
18) 5
19) 12
20) 2
21) 10

1–13 Least Common Multiple

1) 28
2) 15
3) 80
4) 68
5) 24
6) 24

7) 18

8) 30

9) 152

10) 63

11) 551

12) 42

13) 150

14) 8

15) 150

16) 108

17) 72

18) 72

19) 780

20) 120

21) 90

1–14 Divisibility Rules

13) 16   <u>2</u> 3 <u>4</u> 5 6 7 8 9 10

14) 10   <u>2</u> 3 4 <u>5</u> 6 7 8 9 <u>10</u>

15) 15   2 <u>3</u> 4 <u>5</u> 6 7 8 9 10

16) 28   <u>2</u> 3 <u>4</u> 5 6 <u>7</u> 8 9 10

17) 36   <u>2</u> <u>3</u> 4 5 <u>6</u> 7 8 <u>9</u> 10

18) 18   <u>2</u> <u>3</u> 4 5 <u>6</u> 7 8 <u>9</u> 10

19) 27   2 <u>3</u> 4 5 6 7 8 <u>9</u> 10

20) 70   <u>2</u> 3 4 <u>5</u> 6 <u>7</u> 8 9 <u>10</u>

21) 57   2 <u>3</u> 4 5 6 7 8 9 10

22) 102   <u>2</u> <u>3</u> 4 5 <u>6</u> 7 8 9 10

23) 144   <u>2</u> <u>3</u> <u>4</u> 5 <u>6</u> 7 <u>8</u> <u>9</u> 10

24) 75   2 <u>3</u> 4 <u>5</u> 6 7 8 9 10

Chapter 2: Real Numbers and Integers

2–1 Adding and Subtracting Integers

2–2 Multiplying and Dividing Integers

2–3 Ordering Integers and Numbers

2–4 Arrange, Order, and Comparing Integers

2–5 Order of Operations

2–6 Mixed Integer Computations

2–7 Absolute Value

2–8 Integers and Absolute Value

2–9 Classifying Real Numbers Venn Diagram

2–1 Adding and Subtracting Integers

Find the sum.

1) $(-12) + (-4)$

2) $5 + (-24)$

3) $(-14) + 23$

4) $(-8) + (39)$

5) $43 + (-12)$

6) $(-23) + (-4) + 3$

7) $4 + (-12) + (-10) + (-25)$

8) $19 + (-15) + 25 + 11$

9) $(-9) + (-12) + (32 - 14)$

10) $4 + (-30) + (45 - 34)$

Find the difference.

11) $(-14) - (-9) - (18)$

12) $(-9) - (-25)$

13) $(-12) - (8)$

14) $(28) - (-4)$

15) $(34) - (2)$

16) $(55) - (-5) + (-4)$

17) $(9) - (2) - (-5)$

18) $(2) - (4) - (-15)$

19) $(23) - (4) - (-34)$

20) $(-45) - (-87)$

2–2 Multiplying and Dividing Integers

Find each product.

1) $(-8) \times (-2)$

2) 3×6

3) $(-4) \times 5 \times (-6)$

4) $2 \times (-6) \times (-6)$

5) $11 \times (-12)$

6) $10 \times (-5)$

7) 8×8

8) $(-8) \times (-9)$

9) $6 \times (-5) \times 3$

10) $6 \times (-1) \times 2$

Find each quotient.

11) $18 \div 3$

12) $(-24) \div 4$

13) $(-63) \div (-9)$

14) $54 \div 9$

15) $20 \div (-2)$

16) $(-66) \div (-11)$

17) $64 \div 8$

18) $(-121) \div 11$

19) $72 \div 9$

20) $16 \div 4$

2–3 Ordering Integers and Numbers

Order each set of integers from least to greatest.

1) $-15, -19, 20, -4, 1$ ___, ___, ___, ___, ___, ___

2) $6, -5, 4, -3, 2$ ___, ___, ___, ___, ___, ___

3) $15, -42, 19, 0, -22$ ___, ___, ___, ___, ___, ___

4) $26, -91, 0, -13, 67, -55$ ___, ___, ___, ___, ___, ___

5) $-17, -71, 90, -25, -54, -39$ ___, ___, ___, ___, ___, ___

6) $98, 5, 46, 19, 77, 24$ ___, ___, ___, ___, ___, ___

Order each set of integers from greatest to least.

7) $-2, 5, -3, 6, -4$ ___, ___, ___, ___, ___, ___

8) $-37, 7, -17, 27, 47$ ___, ___, ___, ___, ___, ___

9) $32, -27, 19, -17, 15$ ___, ___, ___, ___, ___, ___

10) $68, 81, 21, -18, 94, 72$ ___, ___, ___, ___, ___, ___

2–4 Arrange, Order, and Comparing Integers

Arrange these integers in descending order.

1) 21, 71, – 18, – 10, 82   ___, ___, ___, ___, ___, ___

2) 15, 11, 20, 12, – 9, – 5   ___, ___, ___, ___, ___, ___

3) – 5, 20, 15, 9, –11   ___, ___, ___, ___, ___, ___

4) 19, 18, – 9, – 6, – 11   ___, ___, ___, ___, ___, ___

5) 56, – 34, – 12, – 5, 32   ___, ___, ___, ___, ___, ___

Compare. Use >, =, <

6) – 8 ____ 12   11) – 56 ____ – 58

7) – 10 ____ –16   12) 78 ____ 87

8) 43 ____ 34   13) – 92 ____ – 102

9) 15 ____ –16   14) – 12 ____ – 12

10) – 354 ____ –345   15) – 721 ____ – 821

2–5 Order of Operations

Evaluate each expression.

1) $(2 \times 2) + 5$

2) $24 - (3 \times 3)$

3) $(6 \times 4) + 8$

4) $25 - (4 \times 2)$

5) $(6 \times 5) + 3$

6) $64 - (2 \times 4)$

7) $25 + (1 \times 8)$

8) $(6 \times 7) + 7$

9) $48 \div (4 + 4)$

10) $(7 + 11) \div (-2)$

11) $9 + (2 \times 5) + 10$

12) $(5 + 8) \times \frac{3}{5} + 2$

13) $2 \times 7 - (\frac{10}{9-4})$

14) $(12 + 2 - 5) \times 7 - 1$

15) $(\frac{7}{5-1}) \times (2 + 6) \times 2$

16) $20 \div (4 - (10 - 8))$

17) $\frac{50}{4(5-4)-3}$

18) $2 + (8 \times 2)$

2–6 Mixed Integer Computations

Compute.

1) $(-70) \div (-5)$

2) $(-14) \times 3$

3) $(-4) \times (-15)$

4) $(-65) \div 5$

5) $18 \times (-7)$

6) $(-12) \times (-2)$

7) $\frac{(-60)}{(-20)}$

8) $24 \div (-8)$

9) $22 \div (-11)$

10) $\frac{(-27)}{3}$

11) $4 \times (-4)$

12) $\frac{(-48)}{12}$

13) $(-14) \times (-2)$

14) $(-7) \times (7)$

15) $\frac{-30}{-6}$

16) $(-54) \div 6$

17) $(-60) \div (-5)$

18) $(-7) \times (-12)$

19) $(-14) \times 5$

20) $88 \div (-8)$

2-7 Absolute Value

Evaluate.

1) $|-4| + |-12| - 7$

2) $|-5| + |-13|$

3) $-18 + |-5 + 3| - 8$

4) $|27| \div |9|$

5) $|-9| \div |-1|$

6) $|200| \div |-100|$

7) $|55| \div |11|$

8) $|36| \div |-6|$

9) $|25| \times |-5|$

10) $|-3| \times |-8|$

11) $|12| \times |-5|$

12) $|11| \times |-6|$

13) $|-8| \times |4|$

14) $|-9| \times |-7|$

15) $|43 - 67 + 9| + |-11| - 1$

16) $|-45 + 78| + |23| - |45|$

17) $75 + |-11 - 30| - |2|$

18) $|-3 + 15| + |9 + 4| - 1$

2–8 Integers and Absolute Value

Write absolute value of each number.

1) -4

2) -7

3) -8

4) 4

5) 5

6) -10

7) 1

8) 6

9) 8

10) -2

11) -1

12) 10

13) 3

14) 7

15) -5

16) -3

17) -9

18) 2

19) 4

20) -6

21) 9

Evaluate.

22) $|-43| - |12| + 10$

23) $76 + |-15 - 45| - |3|$

24) $30 + |-62| - 46$

25) $|32| - |-78| + 90$

26) $|-35 + 4| + 6 - 4$

27) $|-4| + |-11|$

28) $|-6 + 3 - 4| + |7 + 7|$

29) $|-9| + |-19| - 5$

2–9 Classifying Real Numbers Venn Diagram

Identify all of the subsets of real number system to which each number belongs:

Example:

0.1259 : Rational number

$\sqrt{2}$: Irrational number

3 : Natural number, whole number, Integer, rational number

1) 0

2) -5

3) -8.5

4) $\sqrt{4}$

5) -10

6) 18

7) 6

8) π

9) $1\frac{2}{7}$

10) -1

11) $\sqrt{5}$

Answers of Worksheets – Chapter 2

2–1 Adding and Subtracting Integers

1) − 16
2) − 19
3) 9
4) 31
5) 31
6) − 24
7) − 43
8) 40
9) − 3
10) − 15
11) − 23
12) 16
13) − 20
14) 32
15) 32
16) 56
17) 12
18) 13
19) 53
20) 42

2–2 Multiplying and Dividing Integers

1) 16
2) 18
3) 120
4) 72
5) − 132
6) − 50
7) 64
8) 72
9) − 90
10) − 12
11) 6
12) − 6
13) 7
14) 6
15) − 10
16) 6
17) 8
18) − 11
19) 8
20) 4

2–3 Ordering Integers and Numbers

1) − 19, − 15, − 4, 1, 20
2) − 5, − 3, 2, 4, 6
3) − 42, − 22, 0, 15, 19
4) − 91, − 55, − 13, 0, 26, 67
5) − 71, − 54, − 39, − 25, − 17, 90
6) 5, 19, 24, 46, 77, 98
7) 6, 5, − 2, − 3, − 4
8) 47, 27, 7, − 17, − 37
9) 32, 19, 15, − 17, − 27
10) 94, 81, 72, 68, 21, − 18

2–4 Arrange and Order, Comparing Integers

1) 82, 71, 21, − 10, − 18

2) 20, 15, 12, 11, − 5, − 9

3) 20, 15, 9, − 5, −11

4) 19, 18, − 6, − 9, − 11

5) 56, 32, − 5, − 12, − 34

6) <   10) <   14) =

7) >   11) >   15) >

8) >   12) <

9) >   13) >

2–5 Order of Operations

1) 9   7) 33   13) 12

2) 15   8) 49   14) 62

3) 32   9) 6   15) 28

4) 17   10) − 9   16) 10

5) 33   11) 29   17) 50

6) 56   12) 9.8   18) 18

2–6 Mixed Integer Computations

1) 14   8) − 3   15) 5

2) − 42   9) − 2   16) − 9

3) 60   10) − 9   17) 12

4) − 13   11) − 16   18) 84

5) − 126   12) − 4   19) − 70

6) 24   13) 28   20) − 11

7) 3   14) − 49

2–7 Absolute Value

1) 9	7) 5	13) 32
2) 18	8) 6	14) 63
3) −24	9) 125	15) 25
4) 3	10) 24	16) 11
5) 9	11) 60	17) 114
6) 2	12) 66	18) 24

2–8 Integers and Absolute Value

1) 4	11) 1	21) 9
2) 7	12) 10	22) 41
3) 8	13) 3	23) 133
4) 4	14) 7	24) 46
5) 5	15) 5	25) 44
6) 10	16) 3	26) 33
7) 1	17) 9	27) 15
8) 6	18) 2	28) 21
9) 8	19) 4	29) 23
10) 2	20) 6	

2–9 Classifying Real Numbers Venn Diagram

1) 0: whole number, integer, rational number
2) −5: integer, rational number
3) −8.5: rational number
4) $\sqrt{4}$: natural number, whole number, integer, rational number
5) −10: integer, rational number
6) 18 : natural number, whole number, integer, rational number
7) 6: natural number, whole number, integer, rational number
8) π: irrational number
9) $1\frac{2}{7}$: rational number
10) −1: integer, rational number
11) $\sqrt{5}$: irrational number

Chapter 3: Proportions and Ratios

3–1 Writing Ratios

3–2 Simplifying Ratios

3–3 Proportional Ratios

3–4 Create a Proportion

3–5 Similar Figures

3–6 Similar Figure Word Problems

3–7 Complete the Ratio Table

3–8 Ratio and Rates Word Problems

3–1 Writing Ratios

Express each ratio as a rate and unite rate.

1) 120 miles on 4 gallons of gas.

2) 24 dollars for 6 books.

3) 200 miles on 14 gallons of gas

4) 24 inches of snow in 8 hours

Express each ratio as a fraction in the simplest form.

5) 3 feet out of 30 feet

6) 18 cakes out of 42 cakes

7) 16 dimes t0 24 dimes

8) 12 dimes out of 48 coins

9) 14 cups to 84 cups

10) 45 gallons to 65 gallons

11) 10 miles out of 40 miles

12) 22 blue cars out of 55 cars

13) 32 pennies to 300 pennies

14) 24 beetles out of 86 insects

3–2 Simplifying Ratios

Reduce each ratio.

1) 21 : 49

2) 20 : 40

3) 10 : 50

4) 14 : 18

5) 45 : 27

6) 49 : 21

7) 100 : 10

8) 12 : 8

9) 35 : 45

10) 8 : 20

11) 25 : 35

12) 21 : 27

13) 52 : 82

14) 12 : 36

15) 24 : 3

16) 15 : 30

17) 3 : 36

18) 8 : 16

19) 6 : 100

20) 2 : 20

21) 10 : 60

22) 14 : 63

23) 68 : 80

24) 8 : 80

3–3 Proportional Ratios

Solve each proportion.

1) $\dfrac{3}{6} = \dfrac{8}{d}$

2) $\dfrac{k}{5} = \dfrac{12}{15}$

3) $\dfrac{30}{5} = \dfrac{12}{x}$

4) $\dfrac{x}{2} = \dfrac{1}{8}$

5) $\dfrac{d}{3} = \dfrac{2}{6}$

6) $\dfrac{27}{7} = \dfrac{30}{x}$

7) $\dfrac{8}{5} = \dfrac{k}{15}$

8) $\dfrac{60}{20} = \dfrac{3}{d}$

9) $\dfrac{x}{3} = \dfrac{12}{18}$

10) $\dfrac{25}{5} = \dfrac{x}{8}$

11) $\dfrac{12}{x} = \dfrac{4}{2}$

12) $\dfrac{x}{4} = \dfrac{18}{2}$

13) $\dfrac{80}{10} = \dfrac{k}{10}$

14) $\dfrac{12}{6} = \dfrac{6}{d}$

15) $\dfrac{x}{4} = \dfrac{30}{5}$

16) $\dfrac{9}{5} = \dfrac{k}{5}$

17) $\dfrac{45}{15} = \dfrac{15}{d}$

18) $\dfrac{60}{x} = \dfrac{10}{3}$

19) $\dfrac{d}{3} = \dfrac{14}{6}$

20) $\dfrac{k}{4} = \dfrac{4}{2}$

21) $\dfrac{4}{2} = \dfrac{x}{7}$

3–4 Create a Proportion

Create proportion from the given set of numbers.

1) 1, 6, 2, 3

2) 12, 144, 1, 12

3) 16, 4, 8, 2

4) 9, 5, 27, 15

5) 7, 10, 60, 42

6) 8, 7, 24, 21

7) 10, 5, 8, 4

8) 3, 12, 8, 2

9) 2, 2, 1, 4

10) 3, 6, 7, 14

11) 2, 6, 5, 15

12) 7, 2, 14, 4

3–5 Similar Figures

Each pair of figures is similar. Find the value of x.

1)

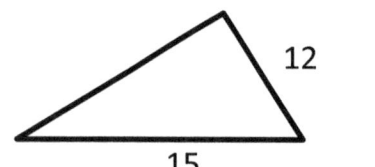 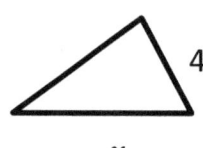

2)

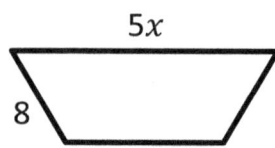 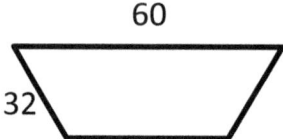

3)

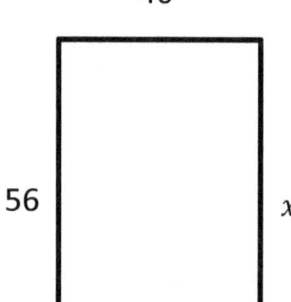 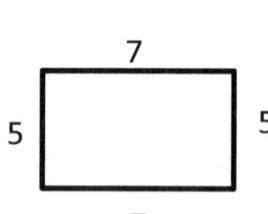

3-6 Similar Figure Word Problems

Answer each question and round your answer to the nearest whole number.

1) If a 42.9 ft tall flagpole casts a 253.1 ft long shadow, then how long is the shadow that a 6.2 ft tall woman casts?

2) A model igloo has a scale of 1 in : 2 ft. If the real igloo is 10 ft wide, then how wide is the model igloo?

3) If a 18 ft tall tree casts a 9 ft long shadow, then how tall is an adult giraffe that casts a 7 ft shadow?

4) Find the distance between San Joe and Mount Pleasant if they are 2 cm apart on a map with a scale of 1 cm : 9 km.

5) A telephone booth that is 8 ft tall casts a shadow that is 4 ft long. Find the height of a lawn ornament that casts a 2 ft shadow.

3–7 Complete the Ratio Table

Complete the ratio tables.

1)

6		18	
7	14		28

2)

3		9	12
7	14		

3)

10		30	40
5	10		20

4)

1	4	5	9
2			

5)

8		40	48
5	10	25	30

6)

2	4	8	10
3			

7)

11		33	44
15	30		

8)

2	6	8	10
5			

9)

4			24
5	10	25	

10)

3	9	21	27
			36

3–8 Ratio and Rates Word Problems

Solve.

1) In a party, 10 soft drinks are required for every 12 guests. If there are 252 guests, how many soft drink is required?

2) In Jack's class, 18 of the students are tall and 10 are short. In Michael's class 54 students are tall and 30 students are short. Which class has a higher ratio of tall to short students?

3) Are these ratios equivalent?
   12 cards to 72 animals, 11 marbles to 66 marbles

4) The price of 3 apples at the Quick Market is $1.44. The price of 5 of the same apples at Walmart is $2.50. Which place is the better buy?

5) The bakers at a Bakery can make 160 bagels in 4 hours. How many bagels can they bake in 16 hours? What is that rate per hour?

6) You can buy 5 cans of green beans at a supermarket for $3.40. How much does it cost to buy 35 cans of green beans?

Answers of Worksheets – Chapter 3

3–1 Writing Ratios

1) $\frac{120\ miles}{4\ gallons}$, 30 miles per gallon

2) $\frac{36\ dollars}{6\ books}$, 6.00 dollars per book

3) $\frac{200\ miles}{14\ gallons}$, 14.29 miles per gallon

4) $\frac{24"\ of\ snow}{8\ hours}$, 4.83 inches of snow per hour

5) $\frac{1}{10}$ 9) $\frac{1}{6}$ 13) $\frac{8}{75}$

6) $\frac{3}{7}$ 10) $\frac{9}{13}$ 14) $\frac{12}{43}$

7) $\frac{2}{3}$ 11) $\frac{1}{4}$

8) $\frac{1}{4}$ 12) $\frac{2}{5}$

3–2 Simplifying Ratios

1) 3 : 7   9) 7 : 9   17) 1 : 12

2) 1 : 2   10) 2 : 5   18) 1 : 2

3) 1 : 5   11) 5 : 7   19) 3 : 50

4) 7 : 9   12) 7 : 9   20) 1 : 10

5) 5 : 3   13) 26 : 41   21) 1 : 6

6) 7 : 3   14) 1 : 3   22) 2 : 9

7) 10 : 1   15) 8 : 1   23) 17 : 20

8) 3 : 2   16) 1 : 2   24) 1 : 10

3–3 Proportional Ratios

1) 16
2) 4
3) 2
4) 0.25
5) 1
6) 7.78
7) 24
8) 1
9) 2
10) 40
11) 6
12) 36
13) 80
14) 3
15) 24
16) 9
17) 5
18) 18
19) 7
20) 8
21) 14

3–4 Create a Proportion

1) 1 : 3 = 2 : 6
2) 12 : 144 = 1 : 12
3) 2 : 4 = 8 : 16
4) 5 : 15 = 9 : 27
5) 7 : 42, 10 : 60
6) 7 : 21 = 8 : 24
7) 8 : 10 = 4 : 5
8) 2 : 3 = 8 : 12
9) 4 : 2 = 2 : 1
10) 7 : 3 = 14 : 6
11) 5 : 2 = 15 : 6
12) 7 : 2 = 14 : 4
13) 6 : 7 = 24 : 28

3–5 Similar Figures

1) 5
2) 3
3) 56

3–6 Similar Figure Word Problems

1) 36.6 ft
2) 5 in
3) 14 ft
4) 18 km
5) 4 ft

3–7 Complete the Ratio Table

1)

6	12	18	24
7	14	21	28

6)

2	4	8	10
3	6	12	15

2)

3	6	9	12
7	14	21	28

7)

11	22	33	44
15	30	45	60

3)

10	20	30	40
5	10	15	20

8)

2	6	8	10
5	15	20	25

4)

1	4	5	9
2	8	10	18

9)

4	8	20	24
5	10	25	30

5)

8	16	40	48
5	10	25	30

10)

3	9	21	27
4	12	28	36

3–8 Ratio and Rates Word Problems

1) 210
2) The ratio for both class is equal to 9 to 5.
3) Yes! Both ratios are 1 to 6.
4) The price at the Quick Market is a better buy.
5) 640, the rate is 40 per hour.
6) $23.80

Chapter 4: Percent

4–1 Converting Between Percent, Fractions, and Decimals

4–2 Table of Common Percent

4–3 Percentage Calculations

4–4 Find What Percentage a Number Is of Another

4–5 Find a Percentage of a Given Number

4–6 Percent Problems

4–7 Percent of Increase and Decrease

4–8 Markup, Discount, and Tax

4–1 Converting Between Percent, Fractions, and Decimals

Converting fractions to decimals.

1) $\dfrac{50}{100}$ 4) $\dfrac{80}{100}$ 7) $\dfrac{90}{100}$

2) $\dfrac{38}{100}$ 5) $\dfrac{7}{100}$ 8) $\dfrac{20}{100}$

3) $\dfrac{15}{100}$ 6) $\dfrac{35}{100}$ 9) $\dfrac{7}{100}$

Write each decimal as a percent.

10) 0.5   13) 0.524   16) 3.63

11) 0.9   14) 0.1   17) 0.008

12) 0.002   15) 0.03   18) 4.78

4–2 Table of Common Percent

Complete the table of common percent.

Fraction	Decimal	Percent
$\frac{1}{25}$	0.04	4%
$\frac{1}{2}$		50%
$\frac{1}{4}$	0.25	25%
$\frac{1}{5}$		20%
$\frac{6}{10}$	0.6	
$\frac{5}{8}$		62.5 %
$\frac{2}{5}$		40%
$\frac{7}{100}$	0.07	7%
$\frac{7}{16}$		43.75%
$\frac{5}{8}$		$62\frac{1}{2}$ %
$\frac{7}{10}$	0.7	
$\frac{30}{100}$		30%
$\frac{4}{8}$	0.8	
$\frac{3}{4}$		75%

4–3 Percentage Calculations

Calculate the percentages.

1) 50% of 25

2) 80% of 15

3) 30% of 34

4) 70% of 45

5) 10% of 0

6) 80% of 22

7) 65% of 8

8) 78% of 54

9) 50% of 80

10) 20% of 10

11) 40% of 40

12) 90% of 0

13) 20% of 70

14) 55% of 60

15) 80% of 10

16) 20% of 880

17) 70% of 100

18) 80% of 90

Solve.

19) 50 is what percentage of 75?

20) What percentage of 100 is 70

21) Find what percentage of 60 is 35.

22) 40 is what percentage of 80?

4–4 Find What Percentage a Number Is of Another

Find the percentage of the numbers.

1) 45 is what percent of 90?

2) 15 is what percent of 75?

3) 20 is what percent of 400?

4) 18 is what percent of 90?

5) 3 is what percent of 15?

6) 8 is what percent of 80?

7) 11 is what percent of 55?

8) 9 is what percent of 90?

9) 2.5 is what percent of 10?

10) 5 is what percent of 25?

11) 60 is what percent of 20?

12) 12 is what percent of 48?

13) 14 is what percent of 28?

14) 8.2 is what percent of 32.8?

15) 1200 is what percent of 4,800?

16) 4,000 is what percent of 20,000?

17) 45 is what percent of 900?

18) 10 is what percent of 200?

19) 15 is what percent of 60?

20) 1.2 is what percent of 24?

4–5 Find a Percentage of a Given Number

Find a Percentage of a Given Number.

1) 90% of 50

2) 40% of 50

3) 10% of 0

4) 80% of 80

5) 60% of 40

6) 50% of 60

7) 30% of 20

8) 35% of 10

9) 10% of 80

10) 10% of 60

11) 100% 0f 50

12) 90% of 34

13) 80% of 42

14) 90% of 12

15) 20% of 56

16) 40% of 40

17) 40% of 6

18) 70% of 38

19) 30% of 3

20) 40% of 50

21) 100% of 8

4–6 Percent Problems

Solve each problem.

1) 51 is 340% of what?

2) 93% of what number is 97?

3) 27% of 142 is what number?

4) What percent of 125 is 29.3?

5) 60 is what percent of 126?

6) 67 is 67% of what?

7) 67 is 13% of what?

8) 41% of 78 is what?

9) 1 is what percent of 52.6?

10) What is 59% of 14 m?

11) What is 90% of 130 inches?

12) 16 inches is 35% of what?

13) 90% of 54.4 hours is what?

14) What percent of 33.5 is 21?

15) Liam scored 22 out of 30 marks in Algebra, 35 out of 40 marks in science and 89 out of 100 marks in mathematics. In which subject his percentage of marks in best?

16) Ella require 50% to pass her test. If she gets 280 marks and falls short by 20 marks, what were the maximum marks she could have got?

4–7 Percent of Increase and Decrease

Find each percent change to the nearest percent, increase or decrease.

1) From 32 grams to 82 grams.

2) From 150 m to 45 m

3) From $438 to $443

4) From 256 ft to 140 ft

5) From 6469 ft to 7488 ft

6) From 36 inches to 90 inches

7) From 54 ft to 104 ft

8) From 84 miles to 24 miles

9) The population of a place in a particular year increased by 15%. Next year it decreased by 15%. Find the net increase or decrease percent in the initial population.

10) The salary of a doctor is increased by 40%. By what percent should the new salary be reduced in order to restore the original salary?

4–8 Markup, Discount, and Tax

Find the selling price of each item.

1) Original price of a microphone: $49.99, discount: 5%, tax: 5%

2) Cost of a pen: $1.95, markup: 70%, discount: 40%, tax: 5%

3) Cost of a puppy: $349.99, markup: 41%, discount: 23%

4) Cost of a shirt: $14.95, markup: 25%, discount: 45%

5) Cost of an oil change: $21.95, markup: 95%

6) Cost of computer: $1,850.00, markup: 75%

Answers of Worksheets – Chapter 4

4–1 Converting Between Percent, Fractions, and Decimals

1) 0.5
2) 0.38
3) 0.15
4) 0.8
5) 0.07
6) 0.35
7) 0.9
8) 0.2
9) 0.07
10) 50%
11) 90%
12) 0.2%
13) 52.4%
14) 10%
15) 3%
16) 363%
17) 0.8%
18) 478%

4–2 Table of Common Percent

Fraction	Decimal	Percent
$\frac{1}{25}$	0.04	4%
$\frac{1}{2}$	0.5	50%
$\frac{1}{4}$	0.25	25%
$\frac{1}{5}$	0.2	20%
$\frac{6}{10}$	0.6	60%
$\frac{5}{8}$	0.625	62.5 %
$\frac{2}{5}$	0.4	40%
$\frac{7}{100}$	0.07	7%
$\frac{7}{16}$	0.4375	43.75%

$\frac{5}{8}$	0.625	625 %
$\frac{7}{10}$	0.7	70%
$\frac{30}{100}$	0.3	30%
$\frac{4}{8}$	0.5	50%
$\frac{3}{4}$	0.75	75%

4–3 Percentage Calculations

1) 12.5
2) 12
3) 10.2
4) 31.5
5) 0
6) 17.6
7) 5.2
8) 42.12
9) 40
10) 2
11) 16
12) 0
13) 14
14) 33
15) 8
16) 176
17) 70
18) 72
19) 67%
20) 70%
21) 58%
22) 50%

4–4 Find What Percentage a Number Is of Another

1) 45 is what percent of 90? 50 %
2) 15 is what percent of 75? 20 %
3) 20 is what percent of 400? 5 %
4) 18 is what percent of 90? 20 %
5) 3 is what percent of 15? 20 %
6) 8 is what percent of 80? 10 %
7) 11 is what percent of 55? 20 %
8) 9 is what percent of 90? 10 %
9) 2.5 is what percent of 10? 25 %
10) 5 is what percent of 25? 20 %
11) 60 is what percent of 20? 300 %
12) 12 is what percent of 48? 25 %
13) 14 is what percent of 28? 50 %
14) 8.2 is what percent of 32.8? 25 %
15) 1200 is what percent of 4,800? 25 %
16) 4,000 is what percent of 20,000? 20 %

17) 45 is what percent of 900? 5 %

18) 10 is what percent of 200? 5 %

19) 15 is what percent of 60? 25 %

20) 1.2 is what percent of 24? 5 %

4–5 Find a Percentage of a Given Number

1) 45
2) 20
3) 0
4) 64
5) 24
6) 30
7) 6
8) 3.5
9) 8
10) 6
11) 50
12) 30.6
13) 33.6
14) 10.8
15) 11.2
16) 16
17) 2.4
18) 26.6
19) 0.9
20) 20
21) 8

4–6 Percent Problems

1) 15
2) 104.3
3) 38.34
4) 23.44%
5) 47.6%
6) 100
7) 515.4
8) 31.98
9) 1.9%
10) 8.3 m
11) 117 inches
12) 45.7 inches
13) 49 hours
14) 62.7%
15) Mathematics
16) 600

4–7 Percent of Increase and Decrease

1) 156.25% increase
2) 70% decrease
3) 1.142% increase
4) 45.31% decrease
5) 15.75% increase
6) 150 % increase
7) 92.6 % increase
8) 71.43% decrease
9) 2.25% decrease
10) $28\frac{4}{7}\%$

4–8 Markup, Discount, and Tax

1) $49.87
2) $2.09
3) $379.98
4) $10.28
5) $36.22
6) $3,237.50

Chapter 5: Algebraic Expressions

5–1 Expressions and Variables

5–2 Simplifying Variable Expressions

5–3 The Distributive Property

5–4 Translate Phrases into an Algebraic Statement

5–5 Evaluating One Variable

5–6 Evaluating Two Variables

5–7 Combining like Terms

5–8 Simplifying Polynomial Expressions

5–1 Expressions and Variables

Simplify each expression.

1) $x + 5x$,

   use $x = 5$

2) $8(-3x + 9) + 6$,

   use $x = 6$

3) $10x - 2x + 6 - 5$,

   use $x = 5$

4) $2x - 3x - 9$,

   use $x = 7$

5) $(-6)(-2x - 4y)$,

   use $x = 1, y = 3$

6) $8x + 2 + 4y$,

   use $x = 9, y = 2$

7) $(-6)(-8x - 9y)$,

   use $x = 5, y = 5$

8) $6x + 5y$,

   use $x = 7, y = 4$

Simplify each expression.

9) $5(-4 + 2x)$

10) $-3 - 5x - 6x + 9$

11) $6x - 3x - 8 + 10$

12) $(-8)(6x - 4) + 12$

13) $9(7x + 4) + 6x$

14) $(-9)(-5x + 2)$

5–2 Simplifying Variable Expressions

Simplify each expression.

1) $-2 - x^2 - 6x^2$

2) $3 + 10x^2 + 2$

3) $8x^2 + 6x + 7x^2$

4) $5x^2 - 12x^2 + 8x$

5) $2x^2 - 2x - x$

6) $(-6)(8x - 4)$

7) $4x + 6(2 - 5x)$

8) $10x + 8(10x - 6)$

9) $9(-2x - 6) - 5$

10) $3(x + 9)$

11) $7x + 3 - 3x$

12) $2.5x^2 \times (-8x)$

Simplify.

13) $-2(4 - 6x) - 3x$, $x = 1$

14) $2x + 8x$, $x = 2$

15) $9 - 2x + 5x + 2$, $x = 5$

16) $5(3x + 7)$, $x = 3$

17) $2(3 - 2x) - 4$, $x = 6$

18) $5x + 3x - 8$, $x = 3$

19) $x - 7x$, $x = 8$

20) $5(-2 - 9x)$, $x = 4$

5–3 The Distributive Property

Use the distributive property to simply each expression.

1) $-(-2-5x)$

2) $(-6x+2)(-1)$

3) $(-5)(x-2)$

4) $-(7-3x)$

5) $8(8+2x)$

6) $2(12+2x)$

7) $(-6x+8)\,4$

8) $(3-6x)(-7)$

9) $(-12)(2x+1)$

10) $(8-2x)\,9$

11) $(-2x)(-1+9x)-4x(4+5x)$

12) $3(-5x-3)+4(6-3x)$

13) $(-2)(x+4)-(2+3x)$

14) $(-4)(3x-2)+6(x+1)$

15) $(-5)(4x-1)+4(x+2)$

16) $(-3)(x+4)-(2+3x)$

5-4 Translate Phrases into an Algebraic Statement

Write an algebraic expression for each phrase.

1) A number increased by forty–two.

2) The sum of fifteen and a number

3) The difference between fifty–six and a number.

4) The quotient of thirty and a number.

5) Twice a number decreased by 25.

6) Four times the sum of a number and − 12.

7) A number divided by − 20.

8) The quotient of 60 and the product of a number and − 5.

9) Ten subtracted from a number.

10) The difference of six and a number.

5–5 Evaluating One Variable

Simplify each algebraic expression.

1) $9 - x$, $x = 3$

2) $x + 2$, $x = 5$

3) $3x + 7$, $x = 6$

4) $x + (-5)$, $x = -2$

5) $3x + 6$, $x = 4$

6) $4x + 6$, $x = -1$

7) $10 + 2x - 6$, $x = 3$

8) $10 - 3x$, $x = 8$

9) $\frac{20}{x} - 3$, $x = 5$

10) $(-3) + \frac{x}{4} + 2x$, $x = 16$

11) $(-2) + \frac{x}{7}$, $x = 21$

12) $(-\frac{14}{x}) - 9 + 4x$, $x = 2$

13) $(-\frac{6}{x}) - 9 + 2x$, $x = 3$

14) $(-2) + \frac{x}{8}$, $x = 16$

15) $8(5x - 12)$, $x = -2$

5–6 Evaluating Two Variables

Simplify each algebraic expression.

1) $2x + 4y - 3 + 2$,

   $x = 5, y = 3$

2) $(-\frac{12}{x}) + 1 + 5y$,

   $x = 6, y = 8$

3) $(-4)(-2a - 2b)$,

   $a = 5, b = 3$

4) $10 + 3x + 7 - 2y$,

   $x = 7, y = 6$

5) $9x + 2 - 4y$,

   $x = 7, y = 5$

6) $6 + 3(-2x - 3y)$,

   $x = 9, y = 7$

7) $12x + y$,

   $x = 4, y = 8$

8) $x \times 4 \div y$,

   $x = 3, y = 2$

9) $2x + 14 + 4y$,

   $x = 6, y = 8$

10) $4a - (5 - b)$,

    $a = 4, b = 6$

5–7 Combining like Terms

Simplify each expression.

1) $5 + 2x - 8$

2) $(-2x + 6)\,2$

3) $7 + 3x + 6x - 4$

4) $(-4) - (3)(5x + 8)$

5) $9x - 7x - 5$

6) $x - 12x$

7) $7(3x + 6) + 2x$

8) $(-11x) - 10x$

9) $3x - 12 - 5x$

10) $13 + 4x - 5$

11) $(-22x) + 8x$

12) $2(4 + 3x) - 7x$

13) $(-4x) - (6 - 14x)$

14) $5(6x - 1) + 12x$

15) $22x + 6 + 2x$

16) $(-13x) - 14x$

17) $(-6x) - 9 + 15x$

18) $(-6x) + 7x$

19) $(-5x) + 12 + 7x$

20) $(-3x) - 9 + 15x$

21) $20x - 19x$

5–8 Simplifying Polynomial Expressions

Simplify each polynomial.

1) $4x^5 - 5x^6 + 15x^5 - 12x^6 + 3x^6$

2) $(-3x^5 + 12 - 4x) + (8x^4 + 5x + 5x^5)$

3) $10x^2 - 5x^4 + 14x^3 - 20x^4 + 15x^3 - 8x^4$

4) $-6x^2 + 5x^2 - 7x^3 + 12 + 22$

5) $12x^5 - 5x^3 + 8x^2 - 8x^5$

6) $5x^3 + 1 + x^2 - 2x - 10x$

7) $14x^2 - 6x^3 - 2x(4x^2 + 2x)$

8) $(4x^4 - 2x) - (4x - 2x^4)$

9) $(3x^2 + 1) - (4 + 2x^2)$

10) $(2x + 2) - (7x + 6)$

11) $(12x^3 + 4x^4) - (2x^4 - 6x^3)$

12) $(12 + 3x^3) + (6x^3 + 6)$

13) $(5x^2 - 3) + (2x^2 - 3x^3)$

14) $(23x^3 - 12x^2) - (2x^2 - 9x^3)$

15) $(4x - 3x^3) - (3x^3 + 4x)$

Answers of Worksheets – Chapter 5

5–1 Expressions and Variables

1) 30
2) −66
3) 41
4) −16
5) 84
6) 82
7) 510
8) 62
9) $10x − 20$
10) $−11x + 6$
11) $3x + 2$
12) $− 48x + 44$
13) $69x + 36$
14) $45x − 18$

5–2 Simplifying Variable Expressions

1) $−7x^2 − 2$
2) $10x^2 + 5$
3) $15x^2 + 6x$
4) $−7x^2 + 8x$
5) $2x^2 − 3x$
6) $−48x + 24$
7) $−26x + 12$
8) $90x − 48$
9) $−18x − 59$
10) $3x + 27$
11) $4x + 3$
12) $−20x^3$
13) 1
14) 20
15) 26
16) 80
17) −22
18) 16
19) −48
20) −190

5–3 The Distributive Property

1) $5x + 2$
2) $6x − 2$
3) $−5x + 10$
4) $3x − 7$
5) $16x + 64$
6) $4x + 24$
7) $−24x + 32$
8) $42x − 21$
9) $−24x − 12$
10) $−18x + 72$
11) $−38x^2 − 14x$
12) $−27x + 15$
13) $−5x − 10$
14) $−6x + 14$
15) $−16x + 13$
16) $−6x − 14$

5–4 Translate Phrases into an Algebraic Statement

1) x + 42
2) 15 + x
3) 56 − x
4) 30/x
5) 2x − 25
6) 4(x + (−12))
7) $\frac{x}{-20}$
8) $\frac{60}{-5x}$
9) x − 10
10) 6 − x

5–5 Evaluating One Variable

1) 6
2) 7
3) 25
4) −7
5) 18
6) 2
7) 10
8) −14
9) 1
10) 33
11) 1
12) −8
13) −5
14) 0
15) −176

5–6 Evaluating Two Variables

1) 21
2) 39
3) 64
4) 26
5) 45
6) −111
7) 56
8) 6
9) 58
10) 17

5–7 Combining like Terms

1) 2x − 3
2) −4x + 12
3) 9x + 3
4) −15x − 28
5) 2x − 5
6) −11x
7) 23x + 42
8) −21x
9) −2x − 12
10) 4x + 8
11) −14x
12) − x + 8
13) 10x − 6
14) 42x − 5
15) 24x + 6
16) −27x
17) 9x − 9
18) x
19) 2x + 12
20) 12x − 9
21) x

5–8 Simplifying Polynomial Expressions

1) $-14x^6 + 19x^5$
2) $2x^5 + 8x^4 + x + 12$
3) $-33x^4 + 29x^3 + 10x^2$
4) $-7x^3 - x^2 + 34$
5) $4x^5 - 5x^3 + 8x^2$
6) $5x^3 + x^2 - 12x + 1$
7) $-14x^3 + 10x^2$
8) $6x^4 - 6x$
9) $x^2 - 3$
10) $-5x - 4$
11) $2x^4 + 18x^3$
12) $9x^3 + 18$
13) $-3x^3 + 7x^2 - 3$
14) $32x^3 - 14x^2$
15) $-6x^3$

Chapter 6: Equations

6–1 One–Step Equations

6–2 One–Step Equation Word Problems

6–3 Two–Step Equations

6–4 Two–Step Equation Word Problems

6–5 Multi–Step Equations

6–6 Absolute Value Equations

6–1 One–Step Equations

Solve each equation.

1) $x + 3 = 17$

2) $22 = (-8) + x$

3) $3x = (-30)$

4) $(-36) = (-6x)$

5) $(-6) = 4 + x$

6) $2 + x = (-2)$

7) $20x = (-220)$

8) $18 = x + 5$

9) $(-23) + x = (-19)$

10) $5x = (-45)$

11) $x - 12 = (-25)$

12) $x - 3 = (-12)$

13) $(-35) = x - 27$

14) $8 = 2x$

15) $(-6x) = 36$

16) $(-55) = (-5x)$

17) $x - 30 = 20$

18) $8x = 32$

19) $36 = (-4x)$

20) $4x = 68$

21) $30x = 300$

6–2 One–Step Equation Word Problems

Solve.

1) How many boxes of envelopes can you buy with $18 if one box costs $3?

2) After paying $6.25 for a salad, Ella has $45.56. How much money did she have before buying the salad?

3) How many packages of diapers can you buy with $50 if one package costs $5?

4) Last week James ran 20 miles more than Michael. James ran 56 miles. How many miles did Michael run?

5) Last Friday Jacob had $32.52. Over the weekend he received some money for cleaning the attic. He now has $44. How much money did he receive?

6) After paying $10.12 for a sandwich, Amelia has $35.50. How much money did she have before buying the sandwich?

6–3 Two–Step Equations

Solve each equation.

1) $5(8 + x) = 20$

2) $(-7)(x - 9) = 42$

3) $(-12)(2x - 3) = (-12)$

4) $6(1 + x) = 12$

5) $12(2x + 4) = 60$

6) $7(3x + 2) = 42$

7) $8(14 + 2x) = (-34)$

8) $(-15)(2x - 4) = 48$

9) $3(x + 5) = 12$

10) $\dfrac{3x - 12}{6} = 4$

11) $(-12) = \dfrac{x + 15}{6}$

12) $110 = (-5)(2x - 6)$

13) $\dfrac{x}{8} - 12 = 4$

14) $20 = 12 + \dfrac{x}{4}$

15) $\dfrac{-24 + x}{6} = (-12)$

16) $(-4)(5 + 2x) = (-100)$

17) $(-12x) + 20 = 32$

18) $\dfrac{-2 + 6x}{4} = (-8)$

19) $\dfrac{x + 6}{5} = (-5)$

20) $(-9) + \dfrac{x}{4} = (-15)$

6–4 Two–Step Equation Word Problems

Solve.

1) The sum of three consecutive even numbers is 48. What is the smallest of these numbers?

2) How old am I if 400 reduced by 2 times my age is 244?

3) For a field trip, 4 students rode in cars and the rest filled nine buses. How many students were in each bus if 472 students were on the trip?

4) The sum of three consecutive numbers is 72. What is the smallest of these numbers?

5) 331 students went on a field trip. Six buses were filled, and 7 students traveled in cars. How many students were in each bus?

6) You bought a magazine for $5 and four erasers. You spent a total of $25. How much did each eraser cost?

6–5 Multi–Step Equations

Solve each equation.

1) $-(2 - 2x) = 10$

2) $-12 = -(2x + 8)$

3) $3x + 15 = (-2x) + 5$

4) $-28 = (-2x) - 12x$

5) $2(1 + 2x) + 2x = -118$

6) $3x - 18 = 22 + x - 3 + x$

7) $12 - 2x = (-32) - x + x$

8) $7 - 3x - 3x = 3 - 3x$

9) $6 + 10x + 3x = (-30) + 4x$

10) $(-3x) - 8(-1 + 5x) = 352$

11) $24 = (-4x) - 8 + 8$

12) $9 = 2x - 7 + 6x$

13) $6(1 + 6x) = 294$

14) $-10 = (-4x) - 6x$

15) $4x - 2 = (-7) + 5x$

16) $5x - 14 = 8x + 4$

17) $40 = -(4x - 8)$

18) $(-18) - 6x = 6(1 + 3x)$

19) $x - 5 = -2(6 + 3x)$

20) $6 = 1 - 2x + 5$

Answers of Worksheets – Chapter 6

6–1 One–Step Equations

1) 14
2) 30
3) − 10
4) 6
5) − 10
6) − 4
7) − 11
8) 13
9) 4
10) − 9
11) − 13
12) − 9
13) − 8
14) 4
15) − 6
16) 11
17) 50
18) 4
19) − 9
20) 17
21) 10

6–2 One–Step Equation Word Problems

1) 6
2) $51.81
3) 10
4) 36
5) 11.48
6) 45.62

6–3 Two–Step Equations

1) − 4
2) 3
3) 2
4) 1
5) 0.5
6) $\frac{4}{3}$
7) $-\frac{73}{8}$
8) $\frac{2}{5}$
9) − 1
10) 12
11) − 87
12) − 8
13) 128
14) 32
15) − 48
16) 10
17) − 1
18) − 5
19) − 31
20) − 24

6–4 Two–Step Equation Word Problems

1) 14
2) 78
3) 52
4) 23
5) 54
6) $5

6–5 Multi–Step Equations

1) 6
2) 2
3) −2
4) 2
5) −20
6) 37
7) 22
8) $\frac{4}{3}$
9) −4
10) −8
11) −6
12) 2
13) 8
14) 1
15) 5
16) −6
17) −8
18) −1
19) −1
20) 0

Chapter 7: Systems of Equations

7–1 Solving Systems of Equations by Graphing

7–2 Solving Systems of Equations by Substitution

7–3 Solving Systems of Equations by Elimination

7–4 Systems of Equations Word Problems

7–1 Solving Systems of Equations by Graphing

Solve each system of equations by graphing.

1) $y = -4x - 2$
   $y = -2x + 1$

2) $y = -8x - 4$
   $y = 2$

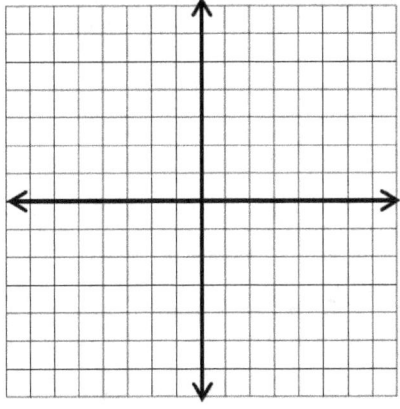
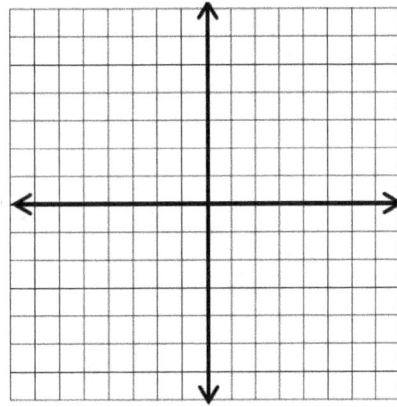

3) $y = 9x - 5$
   $y = -6x + 4$

4) $y = x - 4$
   $y = -2x + 2$

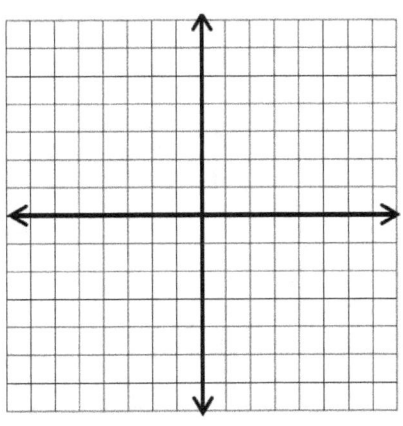
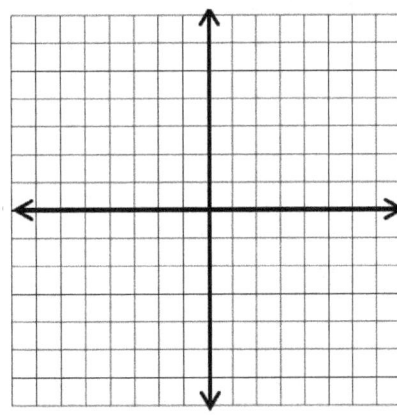

7–2 Solving Systems of Equations by Substitution

Solve each system of equation by substitution.

1) $-x - y = -13$
   $-2x + 2y = 10$

2) $-2x + 2y = 4$
   $-2x + y = 3$

3) $-10x + 2y = -6$
   $6x - 16y = 48$

4) $y = -8$
   $x - 12y = 72$

5) $2y = -6x + 10$
   $10x - 8y = -6$

6) $3x - 9y = -3$
   $3y = 3x - 3$

7) $-4x + 12y = 12$
   $-14x + 16y = -10$

8) $-10x - 16y = 34$
   $4x - 14y = -34$

7–3 Solving Systems of Equations by Elimination

Solve each system of equation by elimination.

1) $x - y = -12$
   $-5x + 3y = 6$

2) $-3x - 4y = 5$
   $x - 2y = 5$

3) $5x - 14y = 22$
   $-6x + 7y = 3$

4) $10x - 14y = -4$
   $-10x - 20y = -30$

5) $32x + 14y = 52$
   $16x - 4y = -40$

6) $2x - 8y = -6$
   $8x + 2y = 10$

7) $-4x + 4y = -4$
   $4x + 2y = 10$

8) $4x + 6y = 10$
   $8x + 12y = -20$

9) $20x - 18y = -12$
   $18x - 8y = 22$

10) $8x + 10y = 52$
    $8x + 6y = 44$

7–4 Systems of Equations Word Problems

1) A farmhouse shelters 10 animals, some are pigs and some are ducks. Altogether there are 36 legs. How many of each animal are there?

2) A class of 195 students went on a field trip. They took vehicles, some cars and some buses. Find the number of cars and the number of buses they took if each car holds 5 students and each bus hold 45 students.

3) The difference of two numbers is 6. Their sum is 14. Find the numbers.

4) The sum of the digits of a certain two–digit number is 7. Reversing its increasing the number by 9. What is the number?

5) The difference of two numbers is 18. Their sum is 66. Find the numbers.

Answers of Worksheets – Chapter 7

7–1 Solving Systems of Equations by Graphing

1)
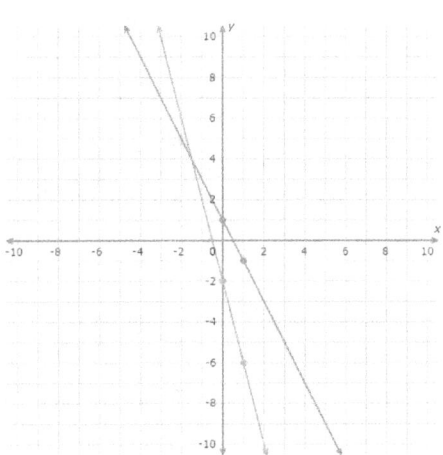

2)
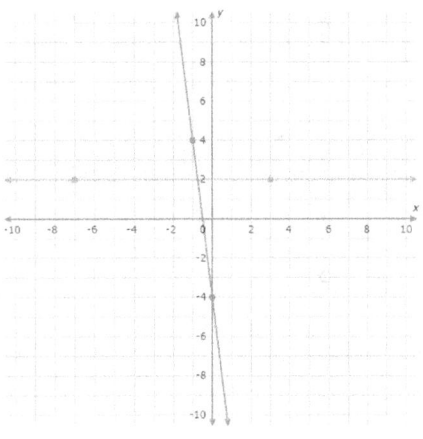

3)
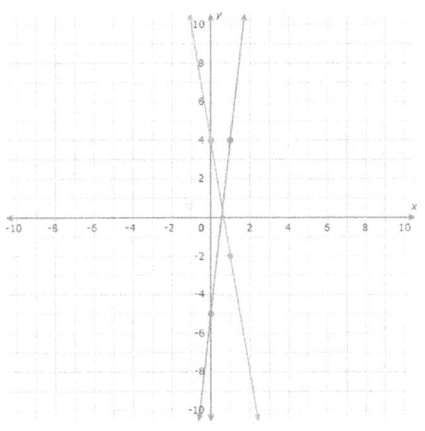

4)
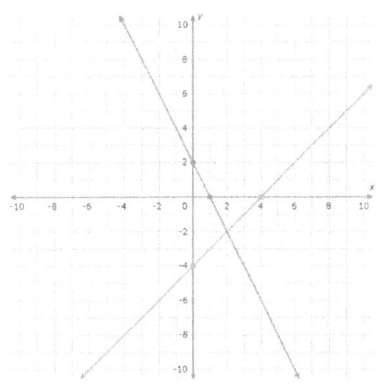

7–2 Solving Systems of Equations by Substitution

1) (4, 9)
2) (−1, 1)
3) (0, −3)
4) (−24, −8)
5) (1, 2)
6) (2, 1)
7) (3, 2)
8) (−5, 1)

7–3 Solving Systems of Equations by Elimination

1) (15, 27)
2) (1, −2)
3) (−4, −3)
4) (1, 1)
5) (−1, 6)
6) (1, 1)
7) (2, 1)
8) No solution
9) (3, 4)
10) (4, 2)

7–4 Systems of Equations Word Problems

1) There are 8 pigs and 2 ducks.
2) There are 3 cars and 4 buses.
3) 10 and 4.
4) 34
5) 24 and 42

Chapter 8: Inequalities

8–1 Graphing Single–Variable Inequalities

8–2 One–Step Inequalities

8–3 Two–Step Inequalities

8–4 Multi–Step Inequalities

8–1 Graphing Single – Variable Inequalities

Draw a graph for each inequality.

1) $-2 > x$

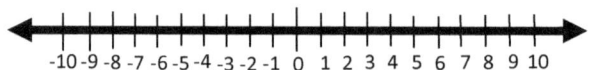

2) $5 \leq -x$

3) $x > 7$

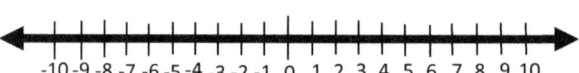

4) $-x > 1.5$

8–2 One–Step Inequalities

Solve each inequality and graph it.

1) $x + 9 \geq 11$

2) $x - 4 \leq 2$

3) $6x \geq 36$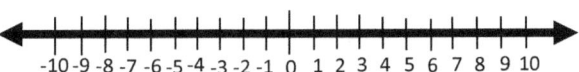

4) $7 + x < 16$

5) $x + 8 \leq 1$

6) $3x > 12$

7) $3x < 24$

8–3 Two–Step Inequalities

Solve each inequality and graph it.

1) $3x - 4 \leq 5$

2) $4x + 19 < 19$

3) $3x + 6 \geq 12$

4) $6x - 5 \geq 19$

5) $2x - 3 < 11$

6) $3 + 4x < 19$

8–4 Multi–Step Inequalities

Solve each inequality and graph it.

1) $\dfrac{7x+1}{3} \geq 5$

2) $\dfrac{9x}{7} - x < 2$

3) $\dfrac{4x+8}{2} \leq 12$

4) $\dfrac{3x-8}{7} > 1$

5) $-3(x-7) > 21$

6) $4 + \dfrac{x}{3} < 7$

Answers of Worksheets – Chapter 8

8–1 Graphing Single–Variable Inequalities

1) $-2 > x$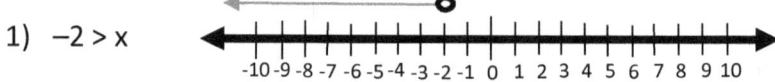

2) $x \leq -5$

3) $x > 7$

4) $-1.5 > x$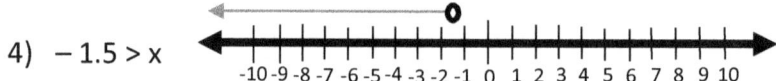

8–2 One–Step Inequalities

1) $x + 9 \geq 11$

2) $x - 4 \leq 2$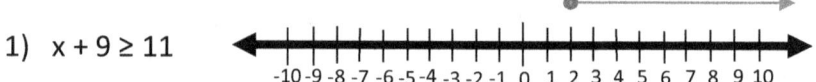

3) $6x \geq 36$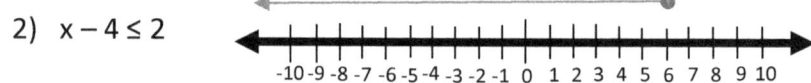

4) $7 + x < 16$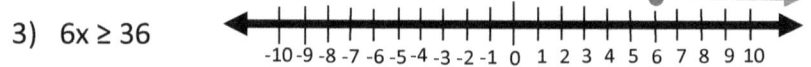

5) $x + 8 \leq 1$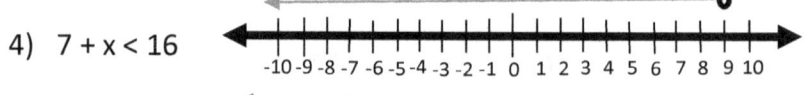

6) $3x > 12$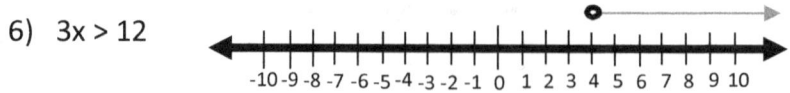

7) $3x < 24$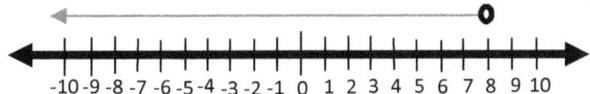

8–3 Two–Step Inequalities

1) $3x - 4 \leq 5$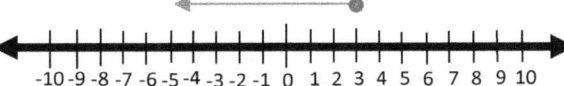

2) $4x + 19 < 19$

3) $3x + 6 \geq 12$

4) $6x - 5 \geq 19$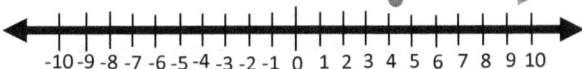

5) $2x - 3 < 11$

6) $3 + 4x < 19$

8–4 Multi–Step Inequalities

1) $\dfrac{7x+1}{3} \geq 5$

2) $\dfrac{9x}{7} - x < 2$

3) $\dfrac{4x+8}{2} \leq 12$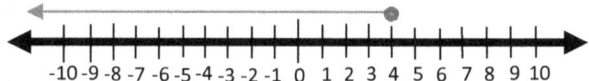

4) $\dfrac{3x-8}{7} > 1$

5) $-3(x-7) > 21$

6) $4 + \dfrac{x}{3} < 7$

Chapter 9: Linear Functions

9-1 Finding Slope

9-2 Graphing Lines Using Slope-Intercept Form

9-3 Graphing Lines Using Standard Form

9-4 Writing Linear Equations

9-5 Graphing Linear Inequalities

9–1 Finding Slope

Find the slope of the line through each pair of points.

1) (2, − 10), (3, 6)

2) (4, − 6), (− 3, − 8)

3) (7, − 12), (5, 10)

4) (19, 3), (20, 3)

5) (15, 8), (− 17, 9)

6) (6, − 12), (15, − 3)

7) (3, 1), (7, − 5)

8) (3, − 2), (− 7, 8)

9) (15, − 3), (− 9, 5)

10) (− 4, 7), (− 6, − 4)

11) (6, − 8), (− 11, − 7)

12) (− 6, 13), (17, − 9)

13) (− 10, − 2), (− 6, − 5)

14) (4, 5), (− 4, 10)

15) (− 3, 1), (− 17, 2)

16) (7, 0), (− 13, − 11)

17) (17, − 13), (17, 8)

18) (12, 2), (− 7, 5)

9–2 Graphing Lines Using Slope–Intercept Form

Sketch the graph of each line.

1) $y = \frac{1}{2}x - 4$

2) $y = -\frac{3}{5}x - 7$

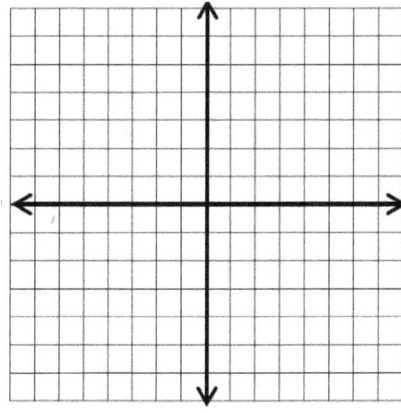
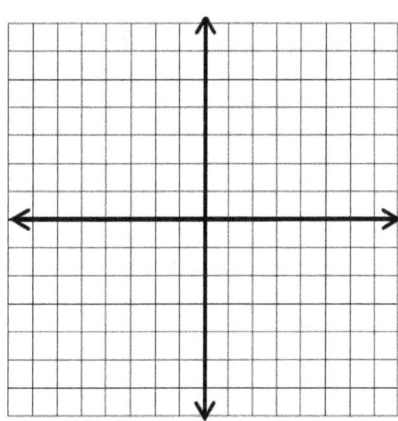

3) $y = \frac{1}{3}x - 8$

5) $y = 6x$

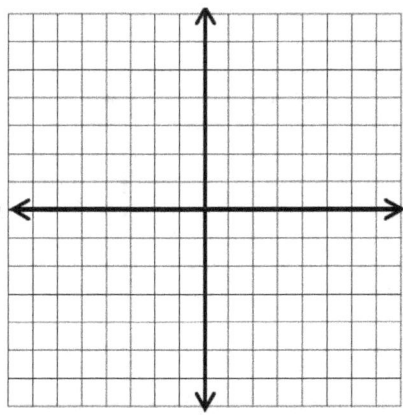
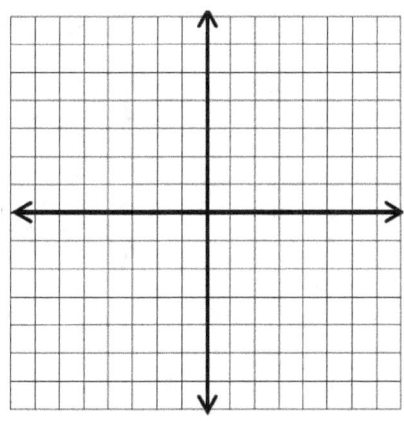

9–3 Graphing Lines Using Standard Form

Sketch the graph of each line.

1) $x + 4y = 12$

2) $2y = -2$

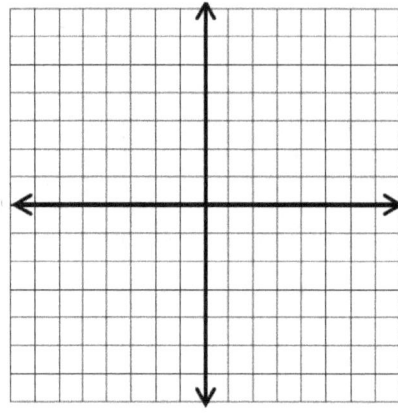
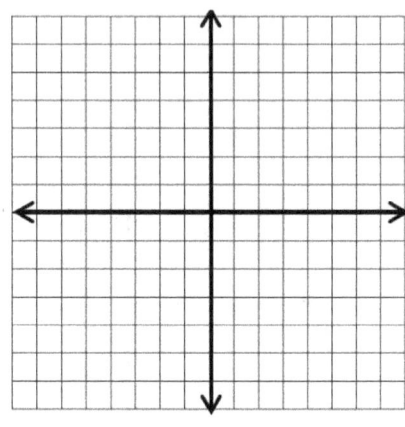

3) $2x - y = 4$

4) $x + y = 2$

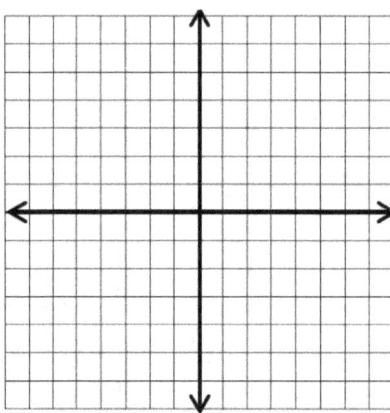
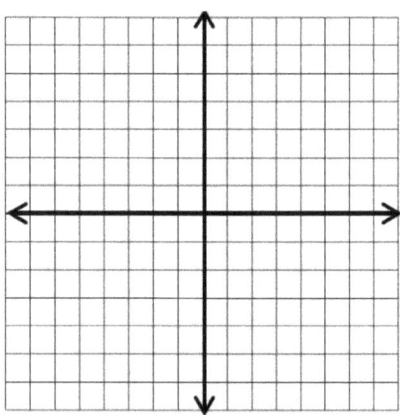

9–4 Writing Linear Equations

Write the slope–intercept form of the equation of the line through the given points.

1) through: $(-4, -2), (-3, 5)$

2) through: $(5, 4), (-4, 3)$

3) through: $(0, -2), (-5, 3)$

4) through: $(-4, -2), (-3, 5)$

5) through: $(0, 3), (-4, -1)$

6) through: $(0, 2), (1, -3)$

7) through: $(0, -5), (4, 3)$

8) through: $(-1, 4), (0, 4)$

9) through: $(2, -3), (3, -5)$

10) through: $(2, 5), (-1, -4)$

11) through: $(1, -3), (-3, 1)$

12) through: $(3, 3), (1, -5)$

13) through: $(4, 4), (3, -5)$

14) through: $(0, 3), (1, 1)$

15) through: $(5, 5), (2, -3)$

16) through: $(-2, -2), (2, -5)$

17) through: $(-3, -2), (1, -1)$

18) through: $(1, 5), (4, 1)$

9–5 Graphing Linear Inequalities

Sketch the graph of each linear inequality.

1) $y < -4x + 2$

2) $2x + y < -4$

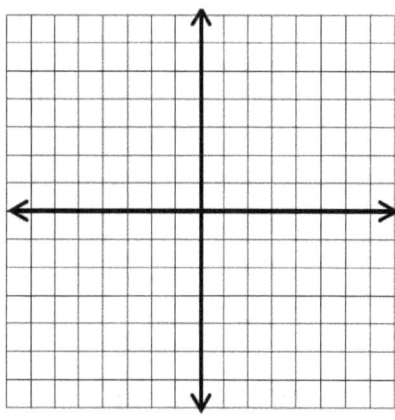
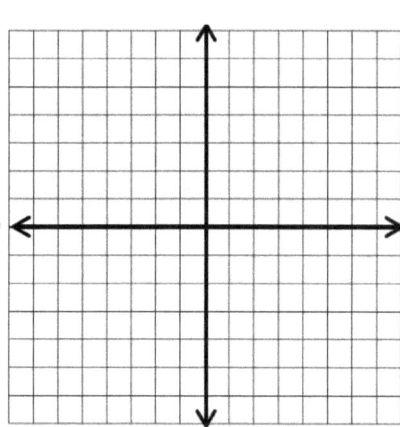

3) $x - 3y < -5$

4) $6x - 2y \geq 8$

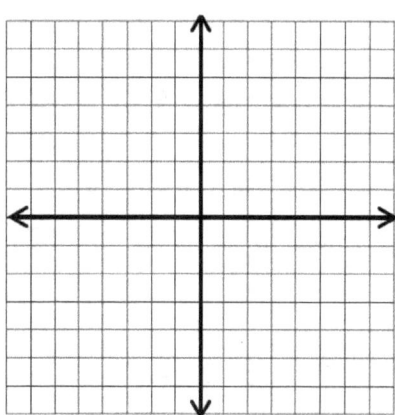
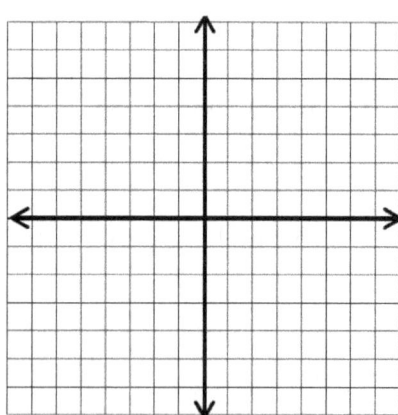

Answers of Worksheets – Chapter 9

9–1 Finding Slope

1) 16
2) $\frac{2}{7}$
3) −11
4) 0
5) $-\frac{1}{32}$
6) 1
7) $-\frac{3}{2}$
8) −1
9) $-\frac{1}{3}$
10) $\frac{11}{2}$
11) $-\frac{1}{17}$
12) $-\frac{22}{23}$
13) $-\frac{3}{4}$
14) $-\frac{5}{8}$
15) $-\frac{1}{14}$
16) $\frac{11}{20}$
17) Undefined
18) $-\frac{3}{19}$

9–2 Graphing Lines Using Slope–Intercept Form

1)
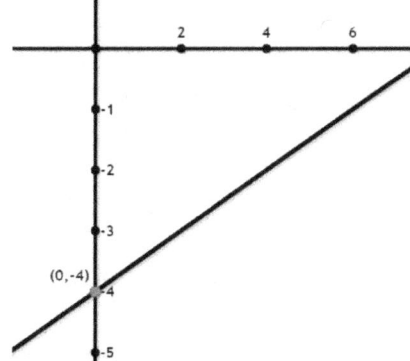

2)
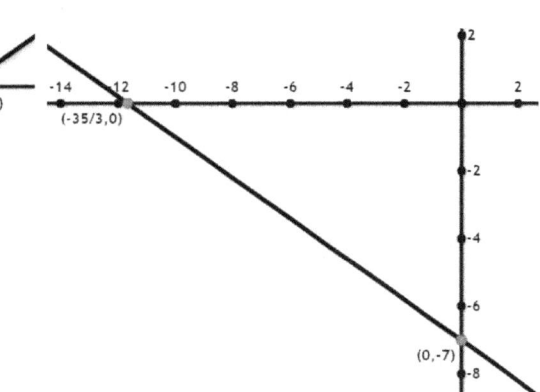

3)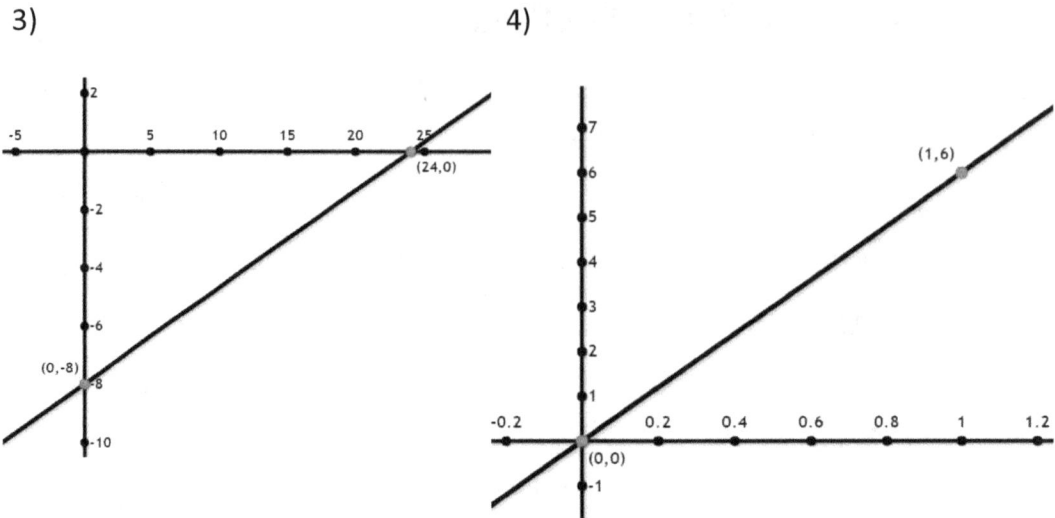

4)

9–3 Graphing Lines Using Standard Form

1)

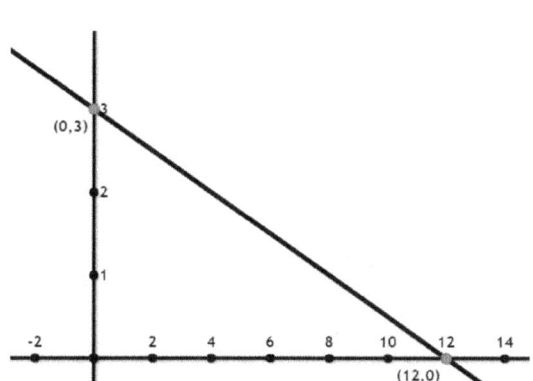

2)

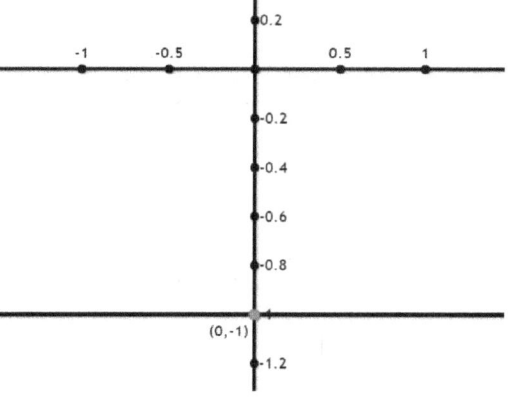

3)

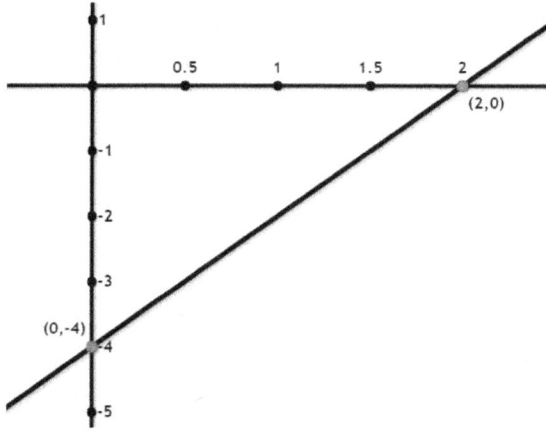

4)

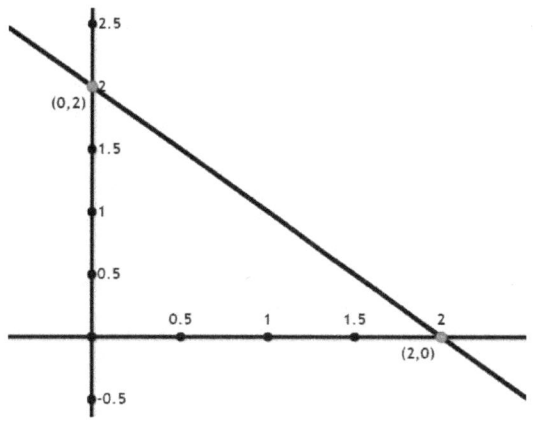

9–4 Writing Linear Equations

1) $y = 7x + 26$
2) $y = \frac{1}{9}x + \frac{31}{9}$
3) $y = -x - 2$
4) $y = -x + 2$
5) $y = x + 3$
6) $y = -5x + 2$
7) $y = 2x - 5$
8) $y = 4$
9) $y = -2x + 1$
10) $y = 3x - 1$
11) $y = -x - 2$
12) $y = 4x - 9$
13) $y = 9x - 32$
14) $y = -2x + 3$
15) $y = \frac{8}{3}x - \frac{25}{3}$
16) $y = -\frac{3}{4}x - \frac{7}{2}$
17) $y = -\frac{1}{4}x - \frac{5}{4}$
18) $y = -\frac{4}{3}x + \frac{19}{3}$

9–5 Graphing Linear Inequalities

1)
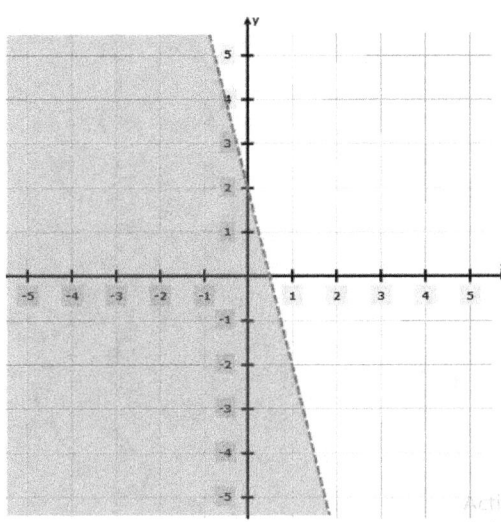

2)
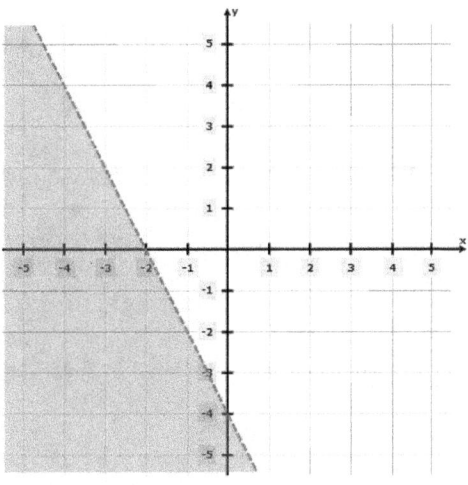

3)
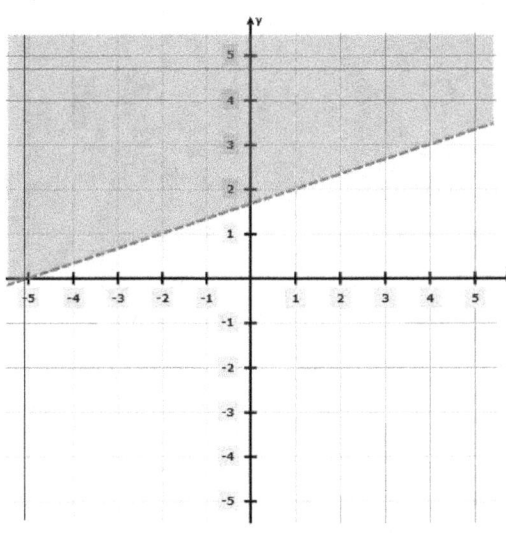

4)
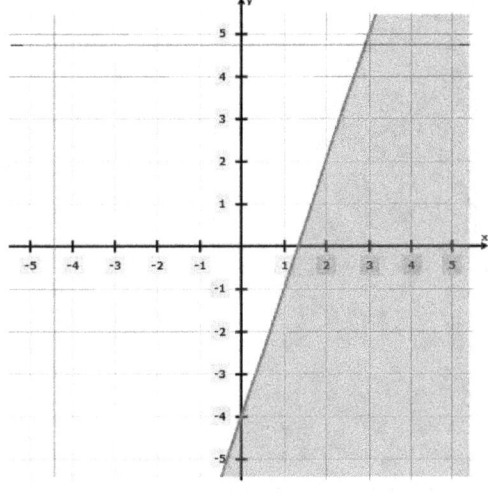

Chapter 10: Exponents and Radicals

11-1 Multiplication Property of Exponents

11-2 Division Property of Exponents

11-3 Powers of Products and Quotients

11-4 Zero and Negative Exponents

11-5 Negative Exponents and Negative Bases

11-6 Writing Scientific Notation

11-7 Square Roots

10–1 Multiplication Property of Exponents

Simplify.

1) x

2) $-5x^4$

3) $7x - 4$

4) -6

5) $8x + 1$

6) $9x^2 - 8x^3$

7) $2x^5$

8) $10 + 8x$

9) $5x^2 - 6x$

10) $-7x^7 + 7x^4$

11) $-8x^4 + 5x^3 - 2x^2 - 8x$

12) $4x - 9x^2 + 4x^3 - 5x^4$

13) $4x^6 + 5x^5 + x^4$

14) $-4 - 2x^2 + 8x$

15) $9x^6 - 8$

16) $7x^5 + 10x^4 - 3x + 10x^7$

17) $4x^6 - 3x^2 - 8x^4$

18) $-5x^4 + 10x - 10$

10–2 Division Property of Exponents

Simplify.

1) $\dfrac{5^5}{5}$

2) $\dfrac{3}{3^5}$

3) $\dfrac{2^2}{2^3}$

4) $\dfrac{2^4}{2^2}$

5) $\dfrac{x}{x^3}$

6) $\dfrac{3x^3}{9x^4}$

7) $\dfrac{2x^{-5}}{9x^{-2}}$

8) $\dfrac{21x^8}{7x^3}$

9) $\dfrac{4x^6}{7x^7}$

10) $\dfrac{4x^2}{6x^3}$

11) $\dfrac{5x}{10x^3}$

12) $\dfrac{2x^3}{3x^5}$

13) $\dfrac{12x^3}{14x^6}$

14) $\dfrac{9x^3}{12y^8}$

15) $\dfrac{25xy^4}{5x^6y^2}$

16) $\dfrac{2x^4}{7x}$

17) $\dfrac{16x^2y^8}{4x^3}$

18) $\dfrac{12x^4}{15x^7y^9}$

19) $\dfrac{12yx^4}{10yx^8}$

20) $\dfrac{16x^4y}{9x^8y^2}$

21) $\dfrac{5x^8}{20x^8}$

10–3 Powers of Products and Quotients

Simplify.

1) $(2x^3)^4$

2) $(4xy^4)^2$

3) $(5x^4)^2$

4) $(11x^5)^2$

5) $(4x^2y^4)^4$

6) $(2x^4y^4)^3$

7) $(3x^2y^2)^2$

8) $(3x^4y^3)^4$

9) $(2x^6y^8)^2$

10) $(12x\ 3x)^3$

11) $(2x^9\ x^6)^3$

12) $(5x^{10}y^3)^3$

13) $(4x^3\ x^2)^2$

14) $(3x^3\ 5x)^2$

15) $(10x^{11}y^3)^2$

16) $(9x^7\ y^5)^2$

17) $(4x^4y^6)^5$

18) $(4x^4)^2$

19) $(3x\ 4y^3)^2$

20) $(9x^2y)^3$

21) $(12x^2y^5)^2$

10–4 Zero and Negative Exponents

Evaluate the following expressions.

1) 8^{-2}

2) 2^{-4}

3) 10^{-2}

4) 5^{-3}

5) 22^{-1}

6) 9^{-1}

7) 3^{-2}

8) 4^{-2}

9) 5^{-2}

10) 35^{-1}

11) 6^{-3}

12) 0^{15}

13) 10^{-9}

14) 3^{-4}

15) 5^{-2}

16) 2^{-3}

17) 3^{-3}

18) 8^{-1}

19) 7^{-3}

20) 6^{-2}

21) $(\frac{2}{3})^{-2}$

22) $(\frac{1}{5})^{-3}$

23) $(\frac{1}{2})^{-8}$

24) $(\frac{2}{5})^{-3}$

10–5 Negative Exponents and Negative Bases

Simplify.

1) -6^{-1}

2) $-4x^{-3}$

3) $-\dfrac{5x}{x^{-3}}$

4) $-\dfrac{a^{-3}}{b^{-2}}$

5) $-\dfrac{5}{x^{-3}}$

6) $\dfrac{7b}{-9c^{-4}}$

7) $-\dfrac{5n^{-2}}{10p^{-3}}$

8) $\dfrac{4ab^{-2}}{-3c^{-2}}$

9) $-12x^2y^{-3}$

10) $\left(-\dfrac{1}{3}\right)^{-2}$

11) $\left(-\dfrac{3}{4}\right)^{-2}$

12) $\left(\dfrac{3a}{2c}\right)^{-2}$

13) $\left(-\dfrac{5x}{3yz}\right)^{-3}$

14) $-\dfrac{2x}{a^{-4}}$

10–6 Writing Scientific Notation

Write each number in scientific notation.

1) 91×10^3

2) 60

3) 2000000

4) 0.0000006

5) 354000

6) 0.000325

7) 2.5

8) 0.00023

9) 56000000

10) 2000000

11) 78000000

12) 0.0000022

13) 0.00012

14) 0.004

15) 78

16) 1600

17) 1450

18) 130000

19) 60

20) 0.113

21) 0.02

10–7 Square Roots

Find the value each square root.

1) $\sqrt{1}$

2) $\sqrt{4}$

3) $\sqrt{9}$

4) $\sqrt{25}$

5) $\sqrt{16}$

6) $\sqrt{49}$

7) $\sqrt{36}$

8) $\sqrt{0}$

9) $\sqrt{64}$

10) $\sqrt{81}$

11) $\sqrt{121}$

12) $\sqrt{225}$

13) $\sqrt{144}$

14) $\sqrt{100}$

15) $\sqrt{256}$

16) $\sqrt{289}$

17) $\sqrt{324}$

18) $\sqrt{400}$

19) $\sqrt{900}$

20) $\sqrt{529}$

21) $\sqrt{90}$

7th Grade PARCC Math Workbook 2018

Answers of Worksheets – Chapter 10

10–1 Multiplication Property of Exponents

1) 4^4
2) 2^5
3) 3^4
4) $3x^4$
5) $36x^5$
6) $12x^3$
7) $25x^8$
8) $36x^5y^4$
9) $63x^3y^8$
10) $28x^4y^7$
11) $4x^4$
12) $24x^7y^6$
13) $560x^{11}y^4$
14) x^{12}
15) $16x^8$
16) x^6
17) $36x^2$
18) $21x^6y^8$

10–2 Division Property of Exponents

1) 5^4
2) $\frac{1}{3^4}$
3) $\frac{1}{2}$
4) 2^2
5) $\frac{1}{x^2}$
6) $\frac{1}{3x}$
7) $\frac{2}{9x^3}$
8) $3x^5$
9) $\frac{4}{7x}$
10) $\frac{2}{3x}$
11) $\frac{1}{2x^2}$
12) $\frac{2}{3x^2}$
13) $\frac{6}{7x^3}$
14) $\frac{3x^3}{4y^8}$
15) $\frac{5y^2}{x^5}$
16) $\frac{2x^3}{7}$
17) $\frac{4y^8}{x}$
18) $\frac{4}{5x^3y^9}$
19) $\frac{6}{5x^4}$
20) $\frac{16}{9x^4y}$
21) $\frac{1}{4}$

10–3 Powers of Products and Quotients

1) $16x^{12}$
2) $16x^2y^8$
3) $25x^8$
4) $121x^{10}$
5) $256x^8y^{16}$
6) $8x^{12}y^{12}$
7) $9x^4y^4$
8) $81x^{16}y^{12}$
9) $4x^{12}y^{16}$
10) $46,656x^6$
11) $8x^{45}$
12) $125x^{30}y^9$

13) $16x^{10}$

14) $225x^8$

15) $100x^{22}y^6$

16) $81x^{14}y^{10}$

17) $1{,}024x^{20}y^{30}$

18) $16x^8$

19) $144x^2y^6$

20) $729x^6y^3$

21) $144x^4y^{10}$

10–4 Zero and Negative Exponents

1) $\frac{1}{64}$

2) $\frac{1}{16}$

3) $\frac{1}{100}$

4) $\frac{1}{125}$

5) $\frac{1}{22}$

6) $\frac{1}{9}$

7) $\frac{1}{9}$

8) $\frac{1}{16}$

9) $\frac{1}{25}$

10) $\frac{1}{35}$

11) $\frac{1}{216}$

12) 0

13) $\frac{1}{1000000000}$

14) $\frac{1}{81}$

15) $\frac{1}{25}$

16) $\frac{1}{8}$

17) $\frac{1}{27}$

18) $\frac{1}{8}$

19) $\frac{1}{343}$

20) $\frac{1}{36}$

21) $\frac{9}{4}$

22) 125

23) 256

24) $\frac{125}{8}$

10–5 Negative Exponents and Negative Bases

1) $-\frac{1}{6}$

2) $-\frac{4}{x^3}$

3) $-5x^4$

4) $-\frac{b^2}{a^3}$

5) $-5x^3$

6) $-\frac{7bc^4}{9}$

7) $-\frac{p^3}{2n^2}$

8) $-\frac{4ac^2}{3b^2}$

9) $-\frac{12x^2}{y^3}$

10) 9

11) $\frac{16}{9}$

12) $\frac{4c^2}{9a^2}$

13) $-\frac{27y^3z^3}{125x^3}$

14) $-2xa^4$

10–6 Writing Scientific Notation

1) 9.1×10^4
2) 6×10^1
3) 2×10^6
4) 6×10^{-7}
5) 3.54×10^5
6) 3.25×10^{-4}
7) 2.5×10^0
8) 2.3×10^{-4}
9) 5.6×10^7
10) 2×10^6
11) 7.8×10^7
12) 2.2×10^{-6}
13) 1.2×10^{-4}
14) 4×10^{-3}
15) 7.8×10^1
16) 1.6×10^3
17) 1.45×10^3
18) 1.3×10^5
19) 6×10^1
20) 1.13×10^{-1}
21) 2×10^{-2}

10–7 Square Roots

1) 1
2) 2
3) 3
4) 5
5) 4
6) 7
7) 6
8) 0
9) 8
10) 9
11) 11
12) 15
13) 12
14) 10
15) 16
16) 17
17) 18
18) 20
19) 30
20) 23
21) $3\sqrt{10}$

Chapter 11: Plane Figures

11–1 Transformations: Translations, Rotations, and Reflections

11–2 The Pythagorean Theorem

11–3 Classifying Triangles and Quadrilaterals

11–4 Area of Triangles

11–5 Perimeter of Polygons

11–6 Area and Circumference of Circles

11–7 Area of Squares, Rectangles, and Parallelograms

11–8 Area of Trapezoids

11–1 Transformations: Translations, Rotations, and Reflections

Graph the image of the figure using the transformation given.

1) translation: 4 units right and 1 unit down

2) translation: 4 units right and 2 unit up

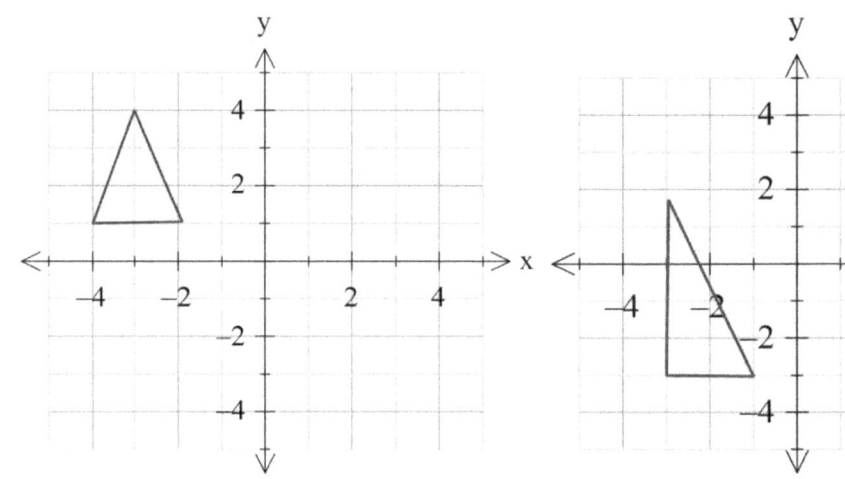

3) rotation 90° counterclockwise about the origin

4) rotation 180° about the origin

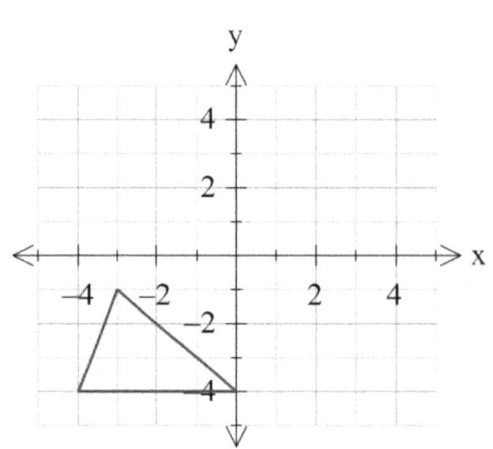
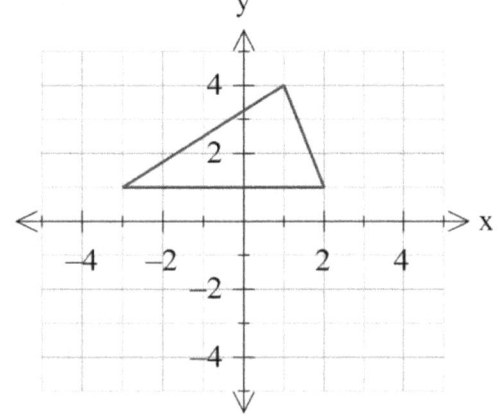

11–2 The Pythagorean Theorem

Do the following lengths form a right triangle?

1)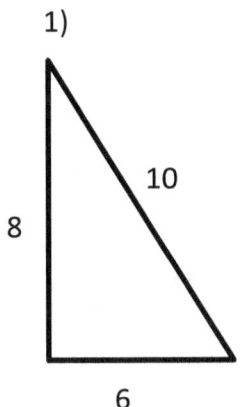

2) 3)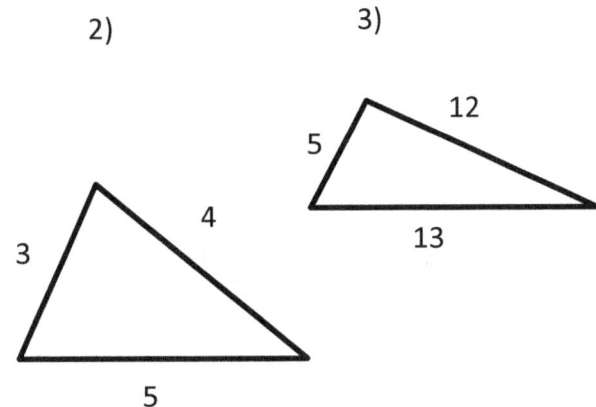

Find each missing length to the nearest tenth.

4)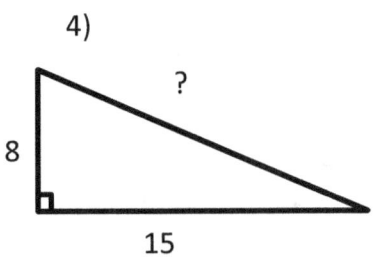

5) 6)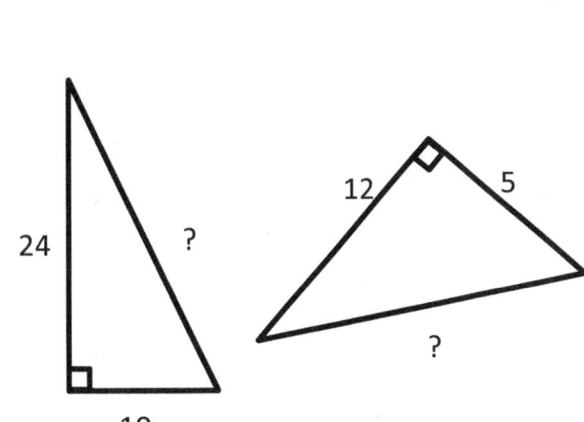

11–3 Classifying Triangles and Quadrilaterals

Classify each triangle by its angles and sides.

1)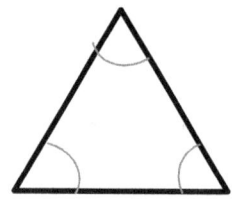

2)

3)

4)

Classify each quadrilateral with the name that best describes it.

5)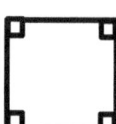

7)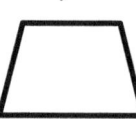

6)

8)

11–4 Area of Triangles

Find the area of each.

1)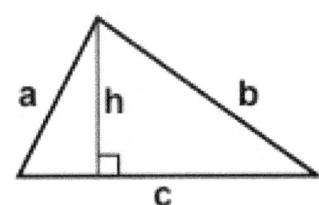

c = 9 mi

h = 3.7 mi

2)

s = 14 m

h = 8 m

3)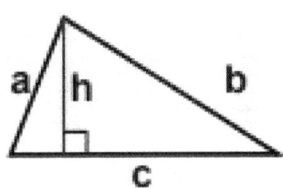

a = 5 m

b = 11 m

c = 14 m

h = 4 m

4)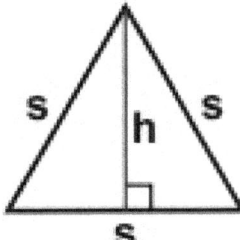

s = 16 m

h = 12.1 m

11–5 Perimeter of Polygons

Find the perimeter of each shape.

1)
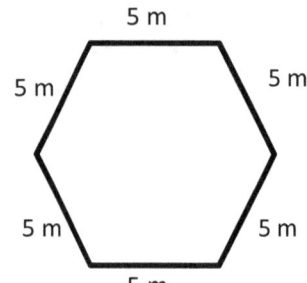

2)
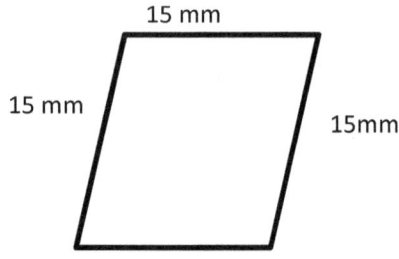

3)
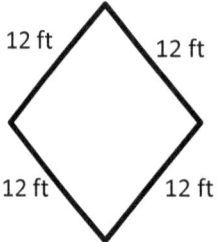

4)
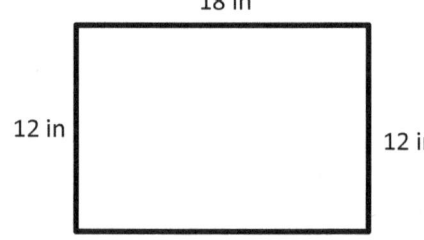

5)
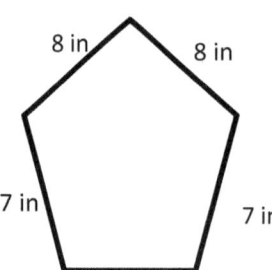

6)
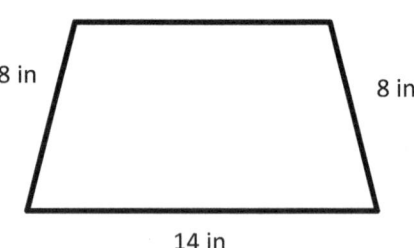

11–6 Area and Circumference of Circles

Find the area and circumference of each.

1)

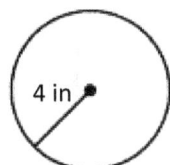

2)

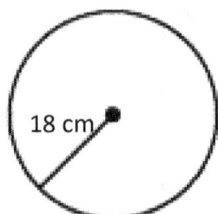

3)

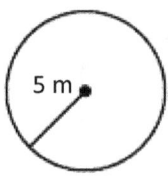

4)

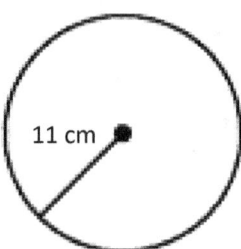

5)

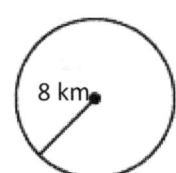

6)

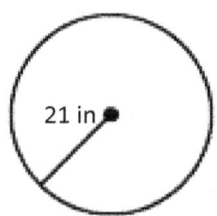

11–7 Area of Squares, Rectangles, and Parallelograms

Find the area of each.

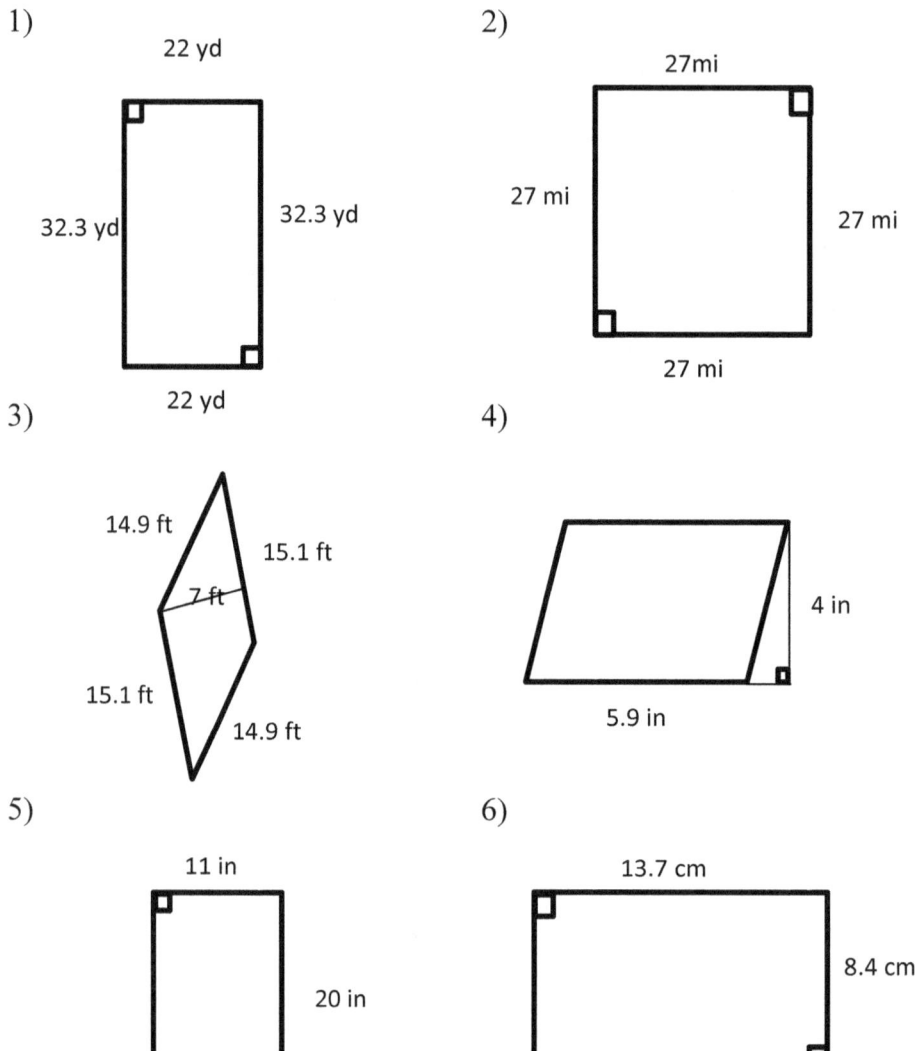

11–8 Area of Trapezoids

Calculate the area for each trapezoid.

1)
2)
3)
4)
5)
6)

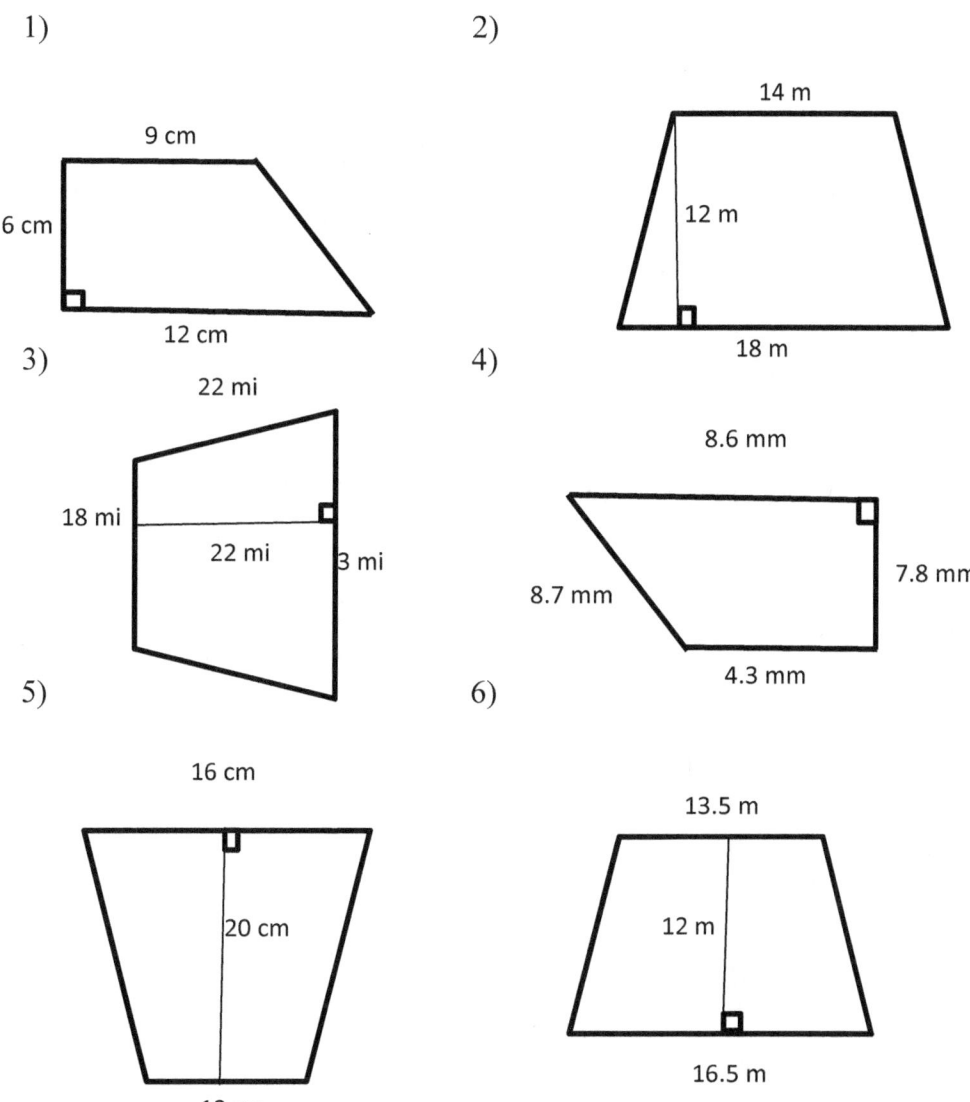

Answers of Worksheets – Chapter 11

11–1 Transformations: Translations, Rotations, and Reflections

1)

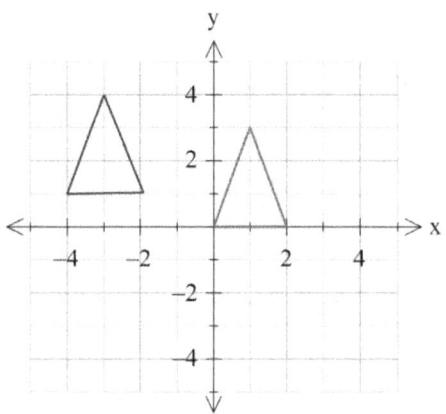

2)

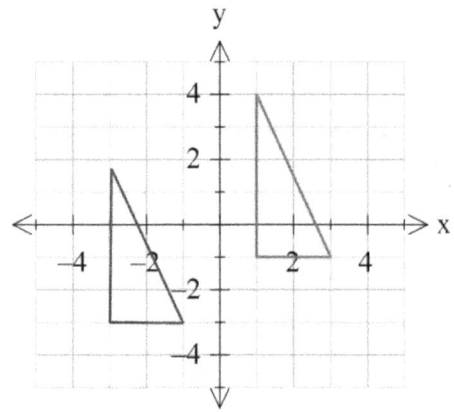

3) rotation 90° counterclockwise about the origin

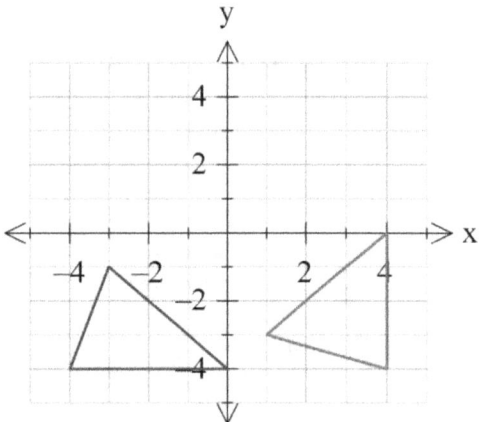

4) rotation 180° about the origin

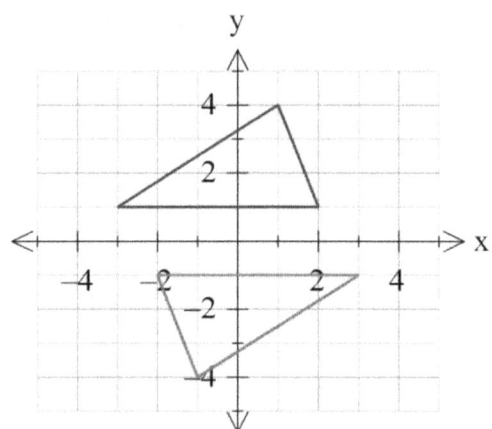

11–2 The Pythagorean Theorem

1) yes
2) yes
3) yes
4) 17
5) 26
6) 13

11–3 Classifying Triangles and Quadrilaterals

1) Right isosceles
2) Equilateral
3) Acute isosceles
4) Acute scalene
5) Rectangle
6) parallelogram
7) Trapezoid
8) Trapezoid

11–4 Area of Triangles

1) 16.65 mi^2
2) 56 m^2
3) 28m^2
4) 96.8 m^2

11–5 Perimeter of Polygons

1) 30 m
2) 60mm
3) 48 ft
4) 60 in
5) 35 in
6) 40 in

11–6 Area and Circumference of Circles

1) Area: 50.27 in^2, Circumference: 25.12 in
2) Area: $1{,}017.36 \text{ cm}^2$, Circumference: 113.04 cm
3) Area: 78.5m^2, Circumference: 31.4 m
4) Area: 379.94 cm^2, Circumference: 69.08 cm
5) Area: 200.96 km^2, Circumference: 50.2 km
6) Area: $1{,}384.74 \text{ km}^2$, Circumference: 131.88 km

11–7 Area of Squares, Rectangles, and Parallelograms

1) 710.6 yd^2
2) 729 mi^2
3) 105.7 ft^2
4) 23.6 in^2
5) 220 in^2
6) 115.08 cm^2

11–8 Area of Trapezoids

1) 63 cm^2
2) 192 m^2
3) 451 mi^2
4) 50.31 nm^2
5) 280 cm^2
6) 180 m^2

Chapter 12: Solid Figures

12-1 Classifying Solids

12-2 Volume of Cubes and Rectangle Prisms

12-3 Surface Area of Cubes

12-4 Surface Area of a Prism

12-1 Classifying Solids

Identify the names of the following shapes.

1)
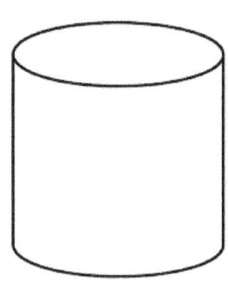

2)
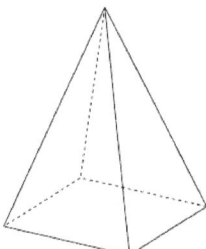

3)

4)
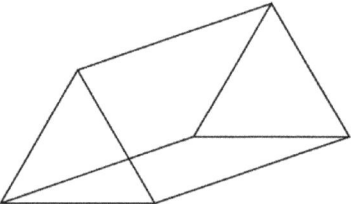

5)
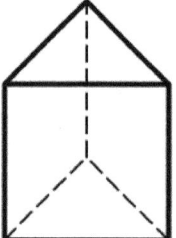

6)

12–2 Volume of Cubes and Rectangle Prisms

Find the volume of each of the rectangular prisms.

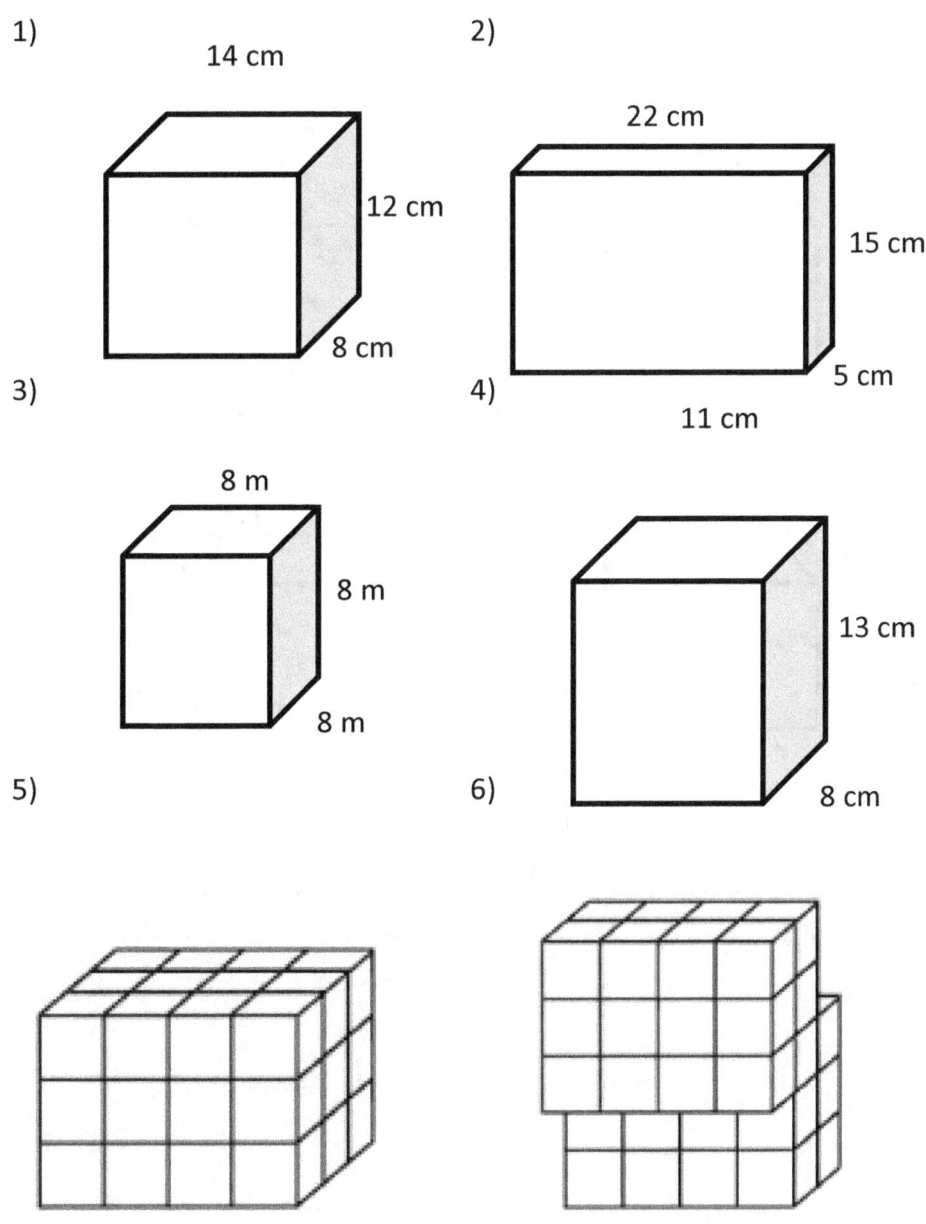

12–3 Surface Area of Cubes

Find the surface of each cube.

1)

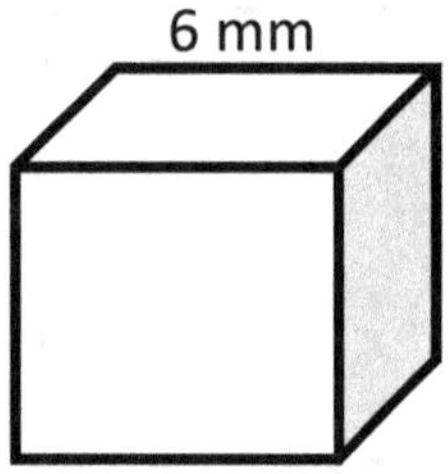

2)

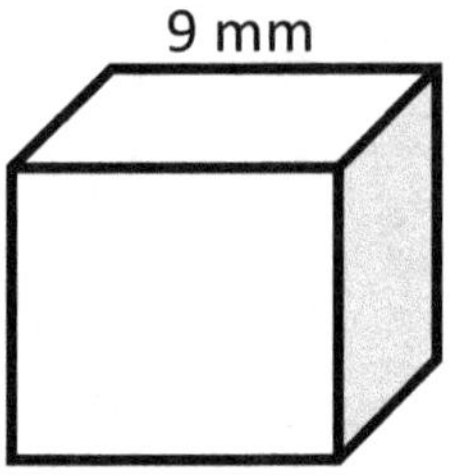

3)

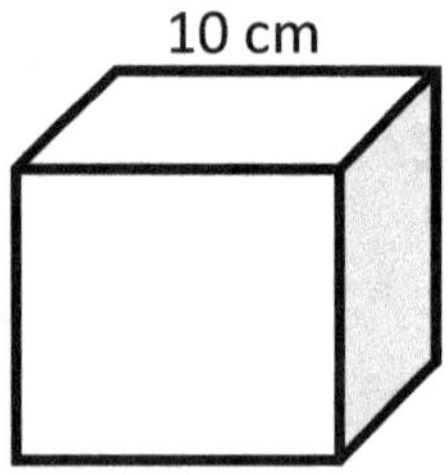

4)

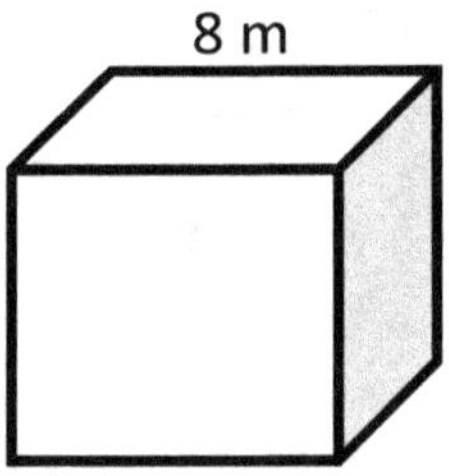

5)

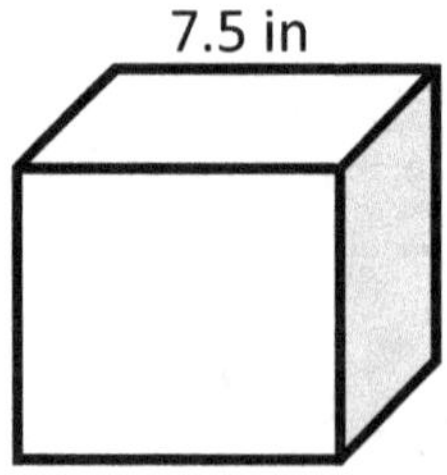

6)

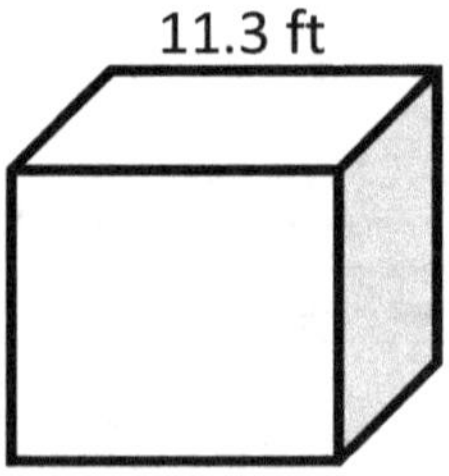

12–4 Surface Area of a Prism

Find the surface of each prism.

1)

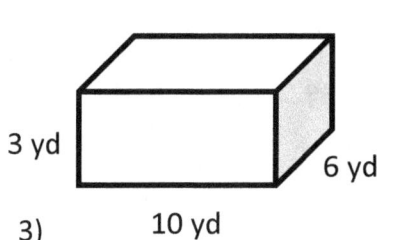

2)

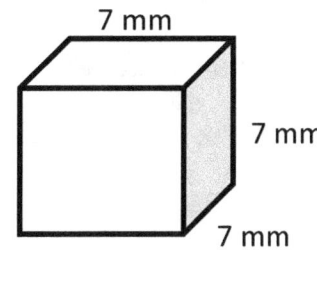

3)

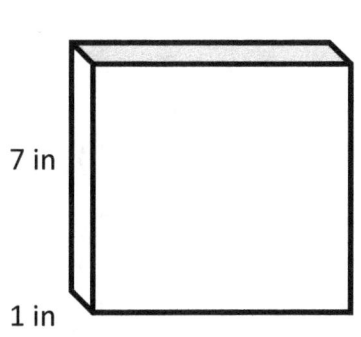

4)

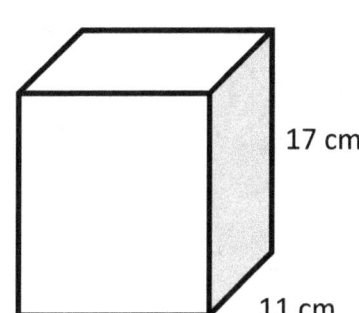

5)

7 in

7 in

1 in

6)

1 cm

3 cm

1 cm

Answers of Worksheets – Chapter 12

12–1 Classifying Solids

1) Cylinder
2) Triangular pyramid
3) Rectangular prism
4) Triangular prism
5) Triangular prism
6) Cylinder

12-2 Volume of Cubes and Rectangle Prisms

1) 1344 cm^3
2) 1650 cm^3
3) 512 m^3
4) 1144 cm^3
5) 36
6) 44

12–3 Surface Area of a Cube

1) 216 mm^2
2) 486 mm^2
3) 600 cm^2
4) 384 m^2
5) 337.5 in^2
6) 766.14 ft^2

12–4 Surface Area of a Prism

1) 216 yd^2
2) 294 mm^2
3) 495.28 in^2
4) 1326 cm^2
5) 126 in^2
6) 14 cm^2

Chapter 13: Statistics

13-1 Mean, Median, Mode, and Range of the Given Data

13-2 First Quartile, Second Quartile and Third Quartile of the Given Data

13-3 Bar Graph

13-4 Box and Whisker Plots

13-5 Stem-And-Leaf Plot

13-6 The Pie Graph or Circle Graph

13-7 Scatter Plots

13–1 Mean, Median, Mode, and Range of the Given Data

Write Mean, Median, Mode, and Range of the Given Data.

1) 7, 2, 5, 1, 1, 2

2) 2, 2, 2, 3, 6, 3, 7, 4

3) 9, 4, 3, 1, 7, 9, 4, 6, 4

4) 8, 4, 2, 4, 3, 2, 4, 5

5) 8, 5, 7, 5, 7, 9, 8

6) 5, 1, 4, 4, 9, 2, 9, 2, 5, 1

7) 4, 1, 5, 9, 7, 7, 5, 4, 3, 5

8) 7, 5, 4, 9, 6, 7, 7, 5, 2

9) 2, 5, 5, 6, 2, 4, 7, 6, 4, 9

10) 10, 5, 2, 5, 4, 5, 8, 10

11) 5, 1, 5, 2, 2

12) 2, 3, 5, 9, 6

13–2 First Quartile, Second Quartile and Third Quartile of the Given Data

Find First Quartile, Second Quartile and Third Quartile of the Given Data.

1) 65, 8, 35, 54, 29, 42, 14, 73, 11

2) 14, 64, 30, 20, 72, 57

3) 99, 37, 83, 62, 74, 49, 59, 40

4) 33, 14, 47, 29, 52, 63, 20, 39, 74, 48

5) 23, 10, 13, 30, 26, 8, 25, 18

6) 35, 60, 20, 80, 95, 15, 40, 85, 75

13–3 Bar Graph

Graph the given information as a bar graph.

Day	Hot dogs sold
Monday	90
Tuesday	70
Wednesday	30
Thursday	20
Friday	60

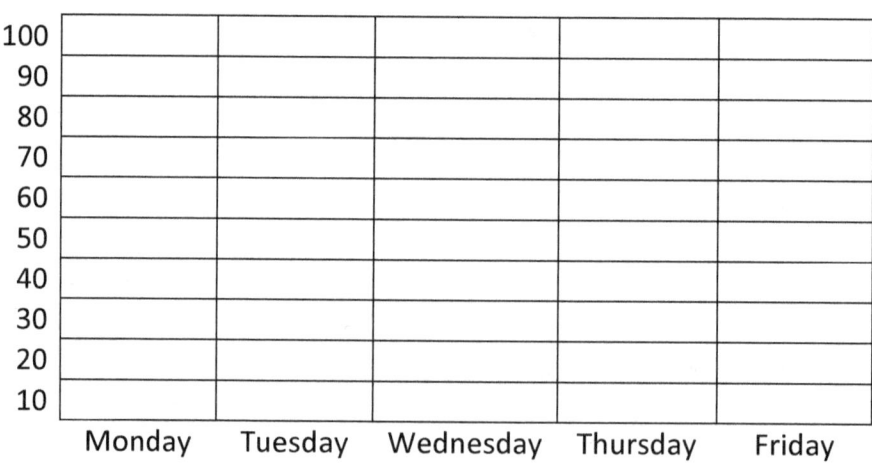

13–4 Box and Whisker Plots

Make box and whisker plots for the given data.

1) 73, 84, 86, 95, 68, 67, 100, 94, 77, 80, 62, 79

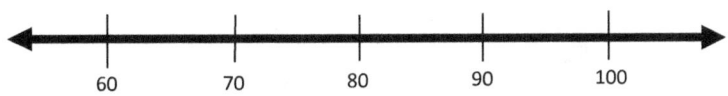

2) 11, 17, 22, 18, 23, 2, 3, 16, 21, 7, 8, 15, 5

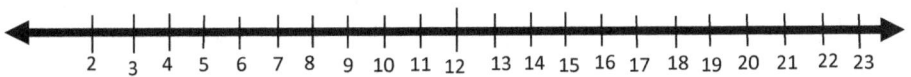

3) 20, 12, 1, 24, 14, 23, 8, 2, 22, 12, 3

13–5 Stem–And–Leaf Plot

Make stem ad leaf plots for the given data.

1) 74, 88, 97, 72, 79, 86, 95, 79, 83, 91

   Key: 8 / 6 =

Stem	Leaf plot

2) 37, 48, 26, 33, 49, 26, 19, 26, 48

   Key: 3 / 7 =

Stem	Leaf plot

3) 58, 41, 42, 67, 54, 65, 65, 54, 69, 53

   Key: 6 / 5 =

Stem	Leaf plot

13–6 The Pie Graph or Circle Graph

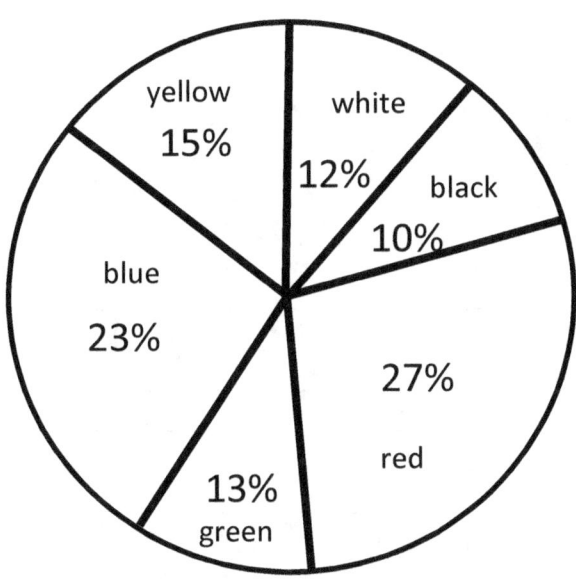

Favorite colors

1) Which color is most?

2) What percentage of pie graph is yellow?

3) Which color is least?

4) What percentage of pie graph is blue?

5) What percentage of pie graph is green?

13–7 Scatter Plots

Construct a scatter plot.

X	Y
1	20
2	40
3	50
4	60

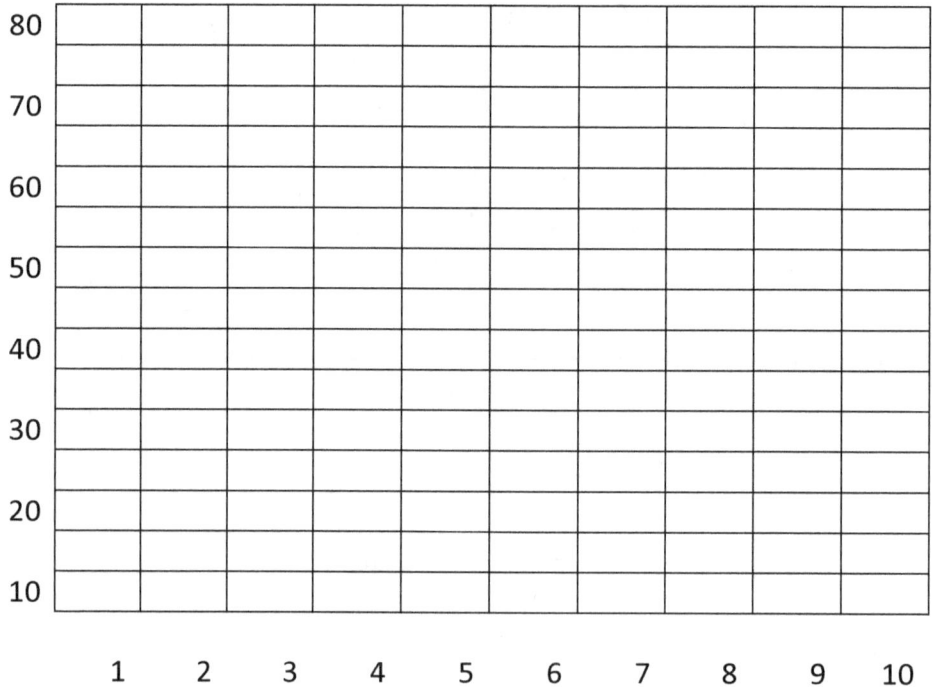

Answers of Worksheets – Chapter 13

13–1 Mean, Median, Mode, and Range of the Given Data

1) mean: 3, median: 2, mode: 2, range: 6
2) mean: 3.625, median: 3, mode: 2, range: 5
3) mean: 5.22, median: 4, mode: 4, range: 8
4) mean: 4, median: 4, mode: 4, range: 6
5) mean: 7, median: 7, mode: 5, 7, 8, range: 4
6) mean: 4.2, median: 4, mode: 1,2,4,5,9, range: 8
7) mean: 5, median: 5, mode: 5, range: 8
8) mean: 5.78, median: 6, mode: 7, range: 7
9) mean: 5, median: 5, mode: 2, 4, 5, 6, range: 7
10) mean: 6.125, median: 5, mode: 5, range: 8
11) mean: 3, median: 2, mode: 2, 5, range: 4
12) mean: 5, median: 5, mode: none, range: 7

13–2 First Quartile, Second Quartile and Third Quartile of the Given Data

1) First quartile: 12.5, second quartile: 35, third quartile: 59.5
2) First quartile: 20, second quartile: 43.5, third quartile: 64
3) First quartile: 44.5, second quartile: 60.5, third quartile: 78.5
4) First quartile: 29, second quartile: 43, third quartile: 52
5) First quartile: 11.5, second quartile: 20.5, third quartile: 25.5
6) First quartile: 27.5, second quartile: 60, third quartile: 82.5

13–3 Bar Graph

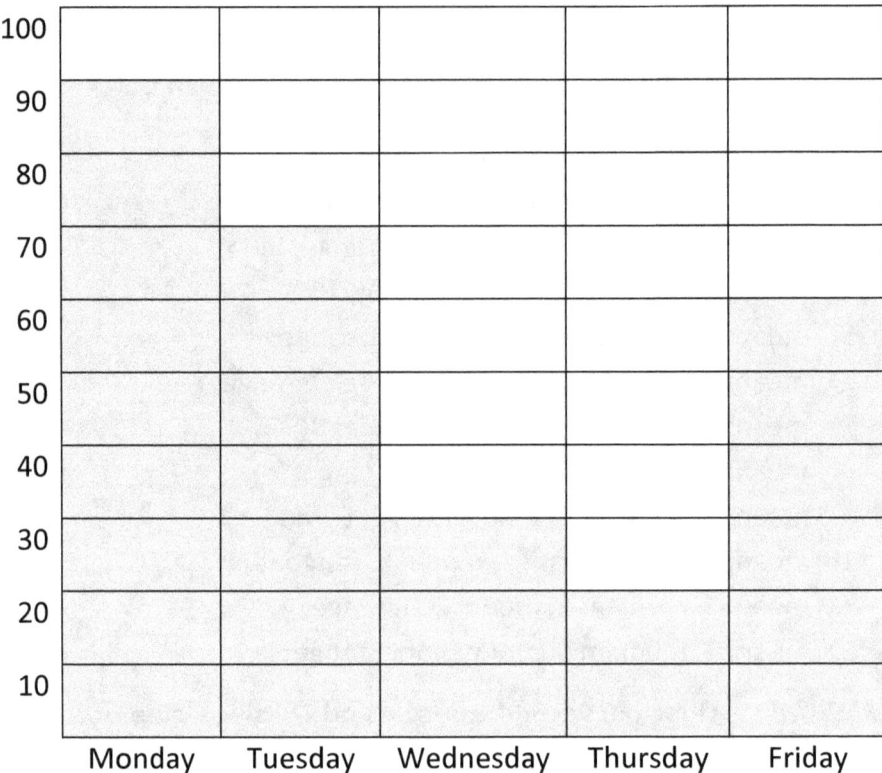

13–4 Box and Whisker Plots

1) 73, 84, 86, 95, 68, 67, 100, 94, 77, 80, 62, 79

Maximum: 100, Minimum: 62, Q_1: 70.5, Q_2: 79.5, Q_3: 90

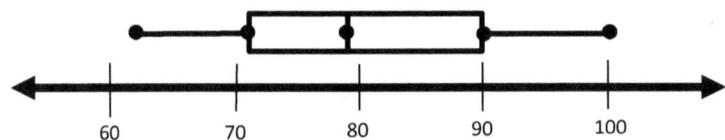

2) 11, 17, 22, 18, 23, 2, 3, 16, 21, 7, 8, 15, 5

Maximum: 23, Minimum: 2, Q_1: 6.5, Q_2: 15.5, Q_3: 19.5

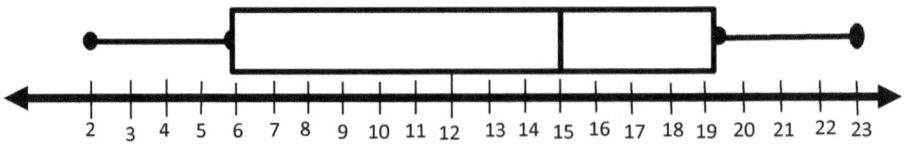

13–5 Stem–And–Leaf Plot

1)

Stem	leaf
7	2 4 9 9
8	3 6 8
9	1 5 7

key: 86

2)

Stem	leaf
1	9
2	6 6 6
3	3 7
4	8 8 9

key:

3)

Stem	leaf
4	1 2
5	3 4 4 8
6	5 5 7 9

key: 65

13–6 The Pie Graph or Circle Graph

1) red
2) 15%
3) black
4) 23%
5) 13%

13–7 Scatter Plots

X	Y
1	20
2	40
3	50
4	60

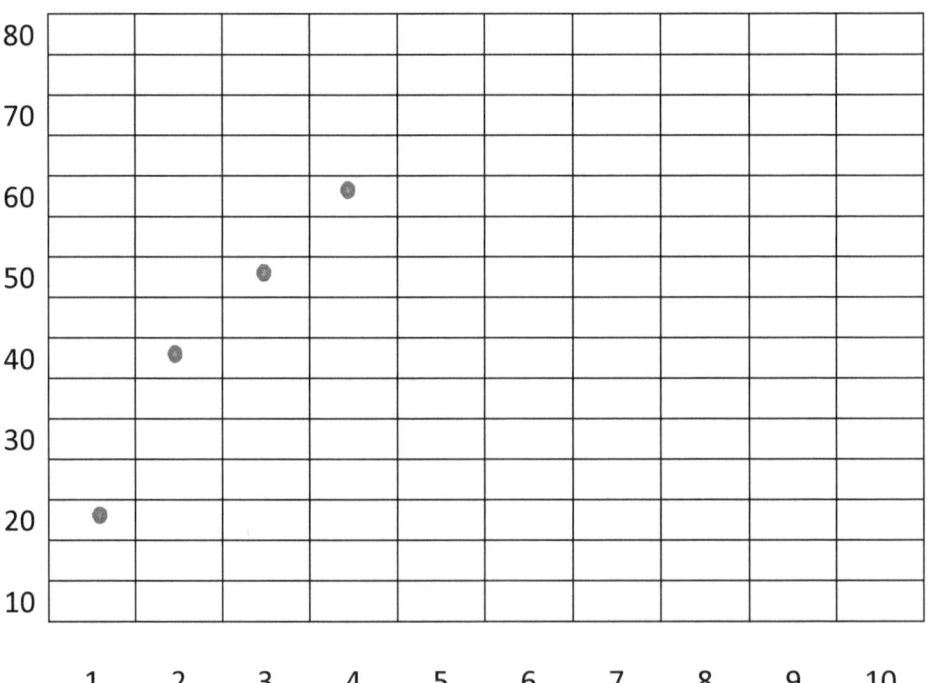

All Mathematics Formulas a 7th grade student Must Know!

Below you will find a list of all Math formulas a 7th grade student MUST have learned before test day, as well as some explanations for how to use them and what they mean. Keep this list around for a quick reminder when your student needs it.

Mathematics Formula Sheet

Place Value
The value of the place, or position, of a digit in a number.
Example: In 456, the 5 is in "tens" position.

Comparing Numbers Signs
Equal to =
Less than <
Greater than >
Greater than or equal ≥
Less than or equal ≤

Rounding
Putting a number up or down to the nearest whole number or the nearest hundred, etc.
Example: 64 rounded to the nearest ten is 60, because 64 is closer to 60 than to 70.

Whole Number
The numbers {0, 1, 2, 3, …}

Estimates
Find a number close to the exact answer.

Decimals
is a fraction written in a special form. For example, instead of writing $\frac{1}{2}$ you can write 0.5.

Fractions
A number expressed in the form $\frac{a}{b}$

Adding and Subtracting with the same denominator:
$$\frac{a}{b} + \frac{c}{b} = \frac{a+c}{b}$$
$$\frac{a}{b} - \frac{c}{b} = \frac{a-c}{b}$$

Adding and Subtracting with the different denominator:
$$\frac{a}{b} + \frac{c}{d} = \frac{ad+cb}{bd}$$
$$\frac{a}{b} - \frac{c}{d} = \frac{ad-cb}{bd}$$

Multiplying and Dividing Fractions:
$$\frac{a}{b} \times \frac{c}{d} = \frac{a \times c}{b \times d}$$

$$\frac{a}{b} \div \frac{c}{d} = \frac{\frac{a}{b}}{\frac{c}{d}} = \frac{ad}{bc}$$

Mixed Numbers
A number composed of a whole number and fraction
Example: $2\frac{2}{3}$
Converting between improper fractions and mixed numbers:
$$a\frac{c}{b} = a + \frac{c}{b} = \frac{ab+c}{b}$$

Factoring Numbers
Factor a number means to break it up into numbers that can be multiplied together to get the original number.
Example: $12 = 2 \times 2 \times 3$

Greatest Common Factor
Multiply common prime factors
Example: $200 = 2 \times 2 \times 2 \times 5 \times 5$
$60 = 2 \times 2 \times 3 \times 5$
GCF (200, 60) = $2 \times 2 \times 5 = 20$

Integers
$\{\ldots, -3, -2, -1, 0, 1, 2, 3, \ldots\}$
Includes: zero, counting numbers, and the negative of the counting numbers

Order of Operations
PEMDAS
(parentheses / exponents / multiply / divide / add / subtract)

Ratios
A ratio is a comparison of two numbers by division.
Example: 3: 5, or $\frac{3}{5}$

Percentages
use the following formula to find part, whole, or percent
part = $\frac{\text{percent}}{100} \times$ whole

Percent of Change
$\frac{\text{New Value} - \text{Old Value}}{\text{Old Value}} \times 100\%$

Markup
Markup = selling price − cost
Markup rate = markup divided by the cost

Divisibility Rules
Divisibility means that you are able to divide a number evenly.
Example: 24 is divisible by 6, because $24 \div 6 = 4$

Least Common Multiple
Check multiples of the largest number
Example: LCM (200, 60): 200 (no), 400 (no), 600 (yes!)

Real Numbers
All numbers that are on number line. Integers plus fractions, decimals, and irrationals ($\sqrt{2}$, $\sqrt{3}$, π, etc.)

Absolute Value
Refers to the distance of a number from 0, the distances are positive as absolute value of a number cannot be negative. $|-22| = 22$

$|x| = \begin{cases} x & \text{for } x \geq 0 \\ -x & \text{for } x < 0 \end{cases}$

$|x| < n \Rightarrow -n < x < n$
$|x| > n \Rightarrow x < -n \text{ or } x > n$

Proportional Ratios
A proportion means that two ratios are equal. It can be written in two ways:
$\frac{a}{b} = \frac{c}{d}$, a: b = c: d

Discount
Multiply the regular price by the rate of discount
Selling price = original price − discount

Expressions and Variables

A variable is a letter that represents unspecified numbers. One may use a variable in the same manner as all other numbers:

Addition	2 + a	2 plus a
Subtraction	y – 3	y minus 3
Division	$\frac{4}{x}$	4 divided by x
Multiplication	5a	5 times a

Systems of Equations

Two or more equations working together.
example:
– 2x + 2y = 4
– 2x + y = 3

Functions

A function is a rule to go from one number (x) to another number (y), usually written y = f (x).
For any given value of x, there can only be one corresponding value y. If y = kx for some number k (example: f (x) = 0.5 x), then y is said to be directly proportional to x. If y = k/x (example: f (x) = 5/x), then y is said to be inversely proportional to x.
The graph of y = f (x h) + k is the translation of the graph of y = f (x) by (h, k) units in the plane. For example, y = f (x + 3) shifts the graph of f (x) by 3 units to the left.

Tax

To find tax, multiply the tax rate to the taxable amount (income, property value, etc.)

Distributive Property

a (b + c) = ab + ac

Polynomial

$P(x) = a_0 x^n + a_1 x^{n-1} + \ldots + a_{n-2} 2x^2 + a_{n-1} x + a_n$

Equations

The values of two mathematical expressions are equal.
ax + b = c

Solving Systems of Equations by Substitution

Consider the system of equations
x – y = 1, –2x + y = 6
Substitute x = 1 – y in the second equation
-2(1-y) + y = 5   y = 2
Substitute y = 2 in x = 1 + y
x = 1 + 2 = 3

Solving Systems of Equations by Elimination

$$x + 2y = 6$$
$$+ \; -x + y = 3$$
$$3y = 9$$
$$y = 3$$

x + 6 = 6
x = 0

Inequalities

Says that two values are not equal

$a \neq b$	a not equal to b
$a < b$	a less than b
$a > b$	a greater than b
$a \geq b$	a greater than or equal b
$a \leq b$	a less than or equal b

Lines (Linear Functions)

Consider the line that goes through points $A(x_1, y_1)$ and $B(x_2, y_2)$.

Distance from A to B:

$$\sqrt{(x_1 - x_2)^2 + (y_1 - y_2)^2}$$

Mid-point of the segment AB:

$$M\left(\frac{x_1+x_2}{2}, \frac{y_1+y_2}{2}\right)$$

Slope of the line:

$$\frac{y_2 - y_1}{x_2 - x_1} = \frac{rise}{run}$$

Point-slope form:

Given the slope m and a point (x_1, y_1) on the line, the equation of the line is $(y - y_1) = m(x - x_1)$.

Slope-intercept form:

given the slope m and the y-intercept b, then the equation of the line is

$y = mx + b$.

Parallel lines

Have equal slopes. Perpendicular lines (i.e., those that make a 90° angle where they intersect) have negative reciprocal slopes: $m_1 \cdot m_2 = -1$.

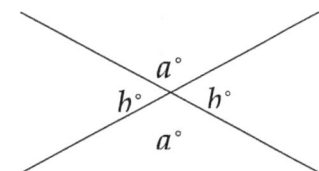

Intersecting Lines

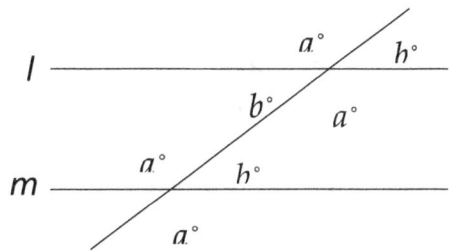

Parallel Lines ($l \parallel m$)

Intersecting lines: opposite angles are equal. Also, each pair of angles along the same line add to 180°. In the figure above, $a + b = 180°$.

Parallel lines: eight angles are formed when a line crosses two parallel lines. The four big angles (a) are equal, and the four small angles (b) are equal.

Parabolas:
A parabola parallel to the y-axis is given by $y = ax^2 + bx + c$.

If $a > 0$, the parabola opens up.
If $a < 0$, the parabola opens down.
The y-intercept is c, and the x-coordinate of the vertex is
$x = -b/2a$.

Exponents
Refers to the number of times a number is multiplied by itself.
$8 = 2 \times 2 \times 2 = 2^3$

Scientific Notation
It is a way of expressing numbers that are too big or too small to be conveniently written in decimal form.
In scientific notation all numbers are written in this form:
$m \times 10^n$

Decimal notation	Scientific notation
5	5×10^0
-25,000	-2.5×10^4
0.5	5×10^{-1}
2,122.456	$2,122456 \times 10^3$

Square
The number we get after multiplying an integer (not a fraction) by itself.
Example: $2 \times 2 = 4$, $2^2 = 4$

Square Roots
A square root of x is a number r whose square is x : $r^2 = x$
r is a square root of x

Factoring
"FOIL"
$(x + a)(x + b) = x^2 + (b + a)x + ab$

"Difference of Squares"
$a^2 - b^2 = (a + b)(a - b)$
$a^2 + 2ab + b^2 = (a + b)(a + b)$
$a^2 - 2ab + b^2 = (a - b)(a - b)$

"Reverse FOIL"
$x^2 + (b + a)x + ab = (x + a)(x + b)$

You can use Reverse FOIL to factor a polynomial by thinking about two numbers a and b which add to the number in front of the x, and which multiply to give the constant. For example, to factor $x^2 + 5x + 6$, the numbers add to 5 and multiply to 6, i.e.:
$a = 2$ and $b = 3$, so that
$x^2 + 5x + 6 = (x + 2)(x + 3)$.

To solve a quadratic such as
$x^2 + bx + c = 0$, first factor the left side to get $(x + a)(x + b) = 0$, then set each part in parentheses equal to zero. For example, $x^2 + 4x + 3 = (x + 3)(x + 1) = 0$ so that $x = -3$ or $x = -1$.
To solve two linear equations in x and y: use the first equation to substitute for a variable in the second. E.g., suppose $x + y = 3$ and $4x - y = 2$. The first equation gives $y = 3 - x$, so the second equation becomes
$4x - (3 - x) = 2 \Rightarrow 5x - 3 = 2$
$\Rightarrow x = 1, y = 2$.

Pythagorean Theorem
$$a^2 + b^2 = c^2$$

Triangles

Right triangles:

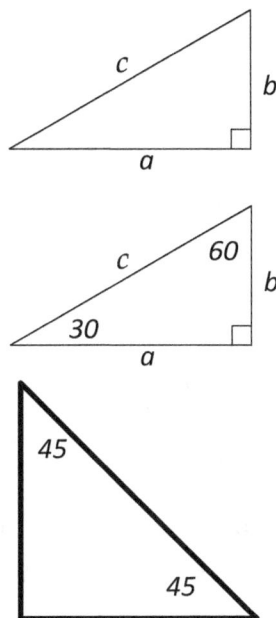

A good example of a right triangle is one with a = 3, b = 4, and c = 5, also called a 3–4–5 right triangle. Note that multiples of these numbers are also right triangles. For example, if you multiply these numbers by 2, you get a = 6, b = 8, and
c = 10 (6–8–10), which is also a right triangle.

All triangles:

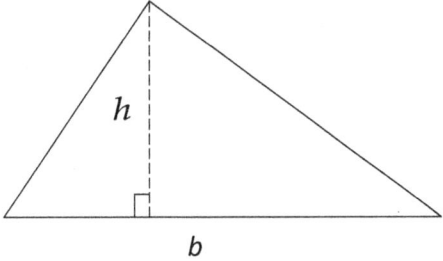

Area = $\frac{1}{2}$ b . h

Angles on the inside of any triangle add up to 180°.
The length of one side of any triangle is always less than the sum and more than the difference of the lengths of the other two sides.
An exterior angle of any triangle is equal to the sum of the two remote interior angles. Other important triangles:

Equilateral:
These triangles have three equal sides, and all three angles are 60°.

Isosceles:
An isosceles triangle has two equal sides. The "base" angles
(the ones opposite the two sides) are equal (see the 45° triangle above).

Similar:
Two or more triangles are similar if they have the same shape. The corresponding angles are equal, and the corresponding sides are in proportion. For example, the 3–4–5 triangle and the 6–8–10 triangle from before are similar since their sides are in a ratio of 2 to 1.

Circles

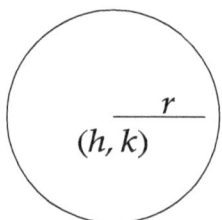

Area = πr²
Circumference = 2πr
Full circle = 360°

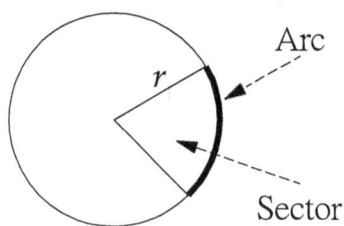

Length Of Arc = (n°/360°) . 2πr
Area Of Sector = (n°/360°) . πr²

Equation of the circle (above left figure): $(x - h)^2 + (y - k)^2 = r^2$.

Area of a parallelogram:

$$A = bh$$

Area of a trapezoid:

$$A = \frac{1}{2}h (b_1 + b_2)$$

Rectangles

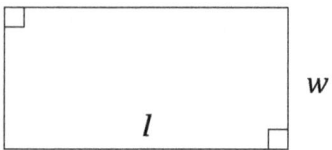

(Square if l = w)
Area = lw

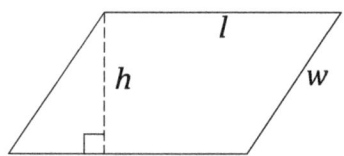

Parallelogram
(Rhombus if l = w)
Area = lh

Regular polygons are n-sided figures with all sides equal and all angles equal.
The sum of the inside angles of an n-sided regular polygon is (n − 2) . 180°.

Surface Area and Volume of a rectangular/right prism:

$$SA = ph + 2B$$
$$V = Bh$$

Solids

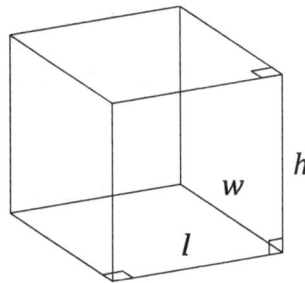

Rectangular Solid
Volume = lwh
Area = 2(lw + wh + lh)

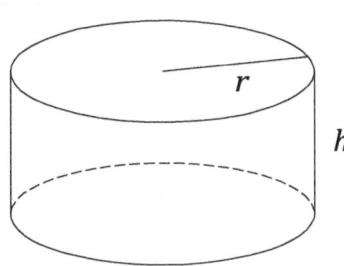

Right Cylinder
Volume = $\pi r^2 h$
Area = $2\pi r(r + h)$

mean: $\dfrac{sum\ of\ the\ data}{of\ data\ entires}$

mode: value in the list that appears most often

range: largest value - smallest value

Surface Area and Volume of a cylinder:
$$SA = 2\pi rh + 2\pi r^2$$
$$V = \pi r^2 h$$

Surface Area and Volume of a Pyramid
$$SA = \frac{1}{2} ps + b$$
$$V = \frac{1}{3} bh$$

Surface Area and Volume of a Cone
$$SA = \pi rs + \pi r^2$$
$$V = \frac{1}{3}\pi r^2 h$$

Surface Area and Volume of a Sphere
$$SA = 4\pi r^2$$
$$V = \frac{4}{3}\pi r^3$$

(p = perimeter of base B; $\pi \sim 3.14$)

Quadratic formula
$$x = \frac{-b \pm \sqrt{b^2 - 4ac}}{2a}$$

Simple interest
$$I = prt$$
(I = interest, p = principal, r = rate, t = time)

Median
Middle value in the list (which must be sorted)
Example: median of
{3, 10, 9, 27, 50} = 10
Example: median of
{3, 9, 10, 27} = $\frac{(9+10)}{2}$ = 9.5

Sum
$$\text{average} \times (\text{number of terms})$$

Average
$$\frac{\text{sum of terms}}{\text{number of terms}}$$

Average speed
$$\frac{\text{total distance}}{\text{total time}}$$

Fundamental Counting Principle:
If an event can happen in N ways, and another, independent event can happen in M ways, then both events together can happen in N × M ways. (Extend this for three or more: $N_1 \times N_2 \times N_3 \ldots$)

Probability
$$\frac{\textit{number of desired outcomes}}{\text{number of total outcomes}}$$

The probability of two different events A and B both happening is:

P(A and B) = p(A) . p(B) as long as the events are independent (not mutually exclusive).

Powers, Exponents, Roots
$x^a \cdot x^b = x^{a+b}$ $x^a/x^b = x^{a-b}$

$1/x^b = x^{-b}$ $(x^a)^b = x^{a \cdot b}$

$(xy)^a = x^a \cdot y^a$

$x^0 = 1$ $\sqrt{xy} = \sqrt{x} \cdot \sqrt{y}$

$(-1)^n = -1$,   if n is odd.
$(-1)^n = +1$,   if n is even.

If $0 < x < 1$, then
$0 < x^3 < x^2 < x < \sqrt{x} < \sqrt[3]{x} < 1$.

Polygon Parts

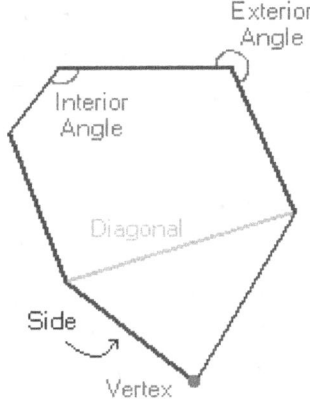

Side
one of the line segments that make up the polygon

Vertex
point where two sides meet. Two or more of these points are called vertices

Diagonal
a line connecting two vertices that isn't a side

Interior Angle
angle formed by two adjacent sides inside the polygon

Exterior Angle
angle formed by two adjacent sides outside the polygon

Factorials

Factorial- the product of a number and all counting numbers below it.

8 factorial = 8! =

8 x 7 x 6 x 5 x 4 x 3 x 2 x 1 = 40,320

5 factorial = 5! =

5 x 4 x 3 x 2 x 1 = 120

2 factorial = 2! = 2 x 1 = 2

Interest

Simple Interest
The charge for borrowing money or the return for lending it.
Interest = principal x rate x time
OR
I=prt

Compound Interest
Interest computed on the accumulated unpaid interest as well as on the original principal.

$$A = P(1+r)^t$$
where
A= amount at end of time
P = principal (starting amount)
r = interest rate (change to a decimal i.e. 50% = .50)
t = number of years invested

Powers/ Exponents

Positive Exponents
An exponent is simply shorthand for multiplying that number of identical factors. So 4^3 is the same as (4)(4)(4), three identical factors of 4. And x^3 is just three factors of x, (x)(x)(x).

Negative Exponents
A negative exponent means to divide by that number of factors instead of multiplying.
So 4^{-3} is the same as $1/(4^3)$, and $x^{-3} = 1/x^3$.

Multiplying Two Powers of the SAME Base
When the bases are the same, you find the new power by just adding the exponents
$$x^a \cdot x^b = x^{(a+b)}$$

Multiplying Two Powers of Different Bases Same Exponent
If the bases are different but the exponents are the same, then you can combine them
$$x^a \cdot y^a = (xy)^a$$

Powers of Powers
For power of a power: you multiply the exponents.
$$(x^a)^b = x^{(ab)}$$

Dividing Powers
$$x^a/x^b = x^a \cdot x^{-b} = x^{a-b}$$

The Zero Exponent
Anything to the 0 power is 1.
$$x^0 = 1$$

Measurements
Metric
millimeter (mm) ~ width of a dime
centimeter (cm) ~ 1 finger sideways
meter (m) ~ height of chair/guitar
kilometer (km) ~ 10 min walk

Customary
inch (in) ~ width of 2 fingers
foot (ft) ~ adult shoe
yard (yd) ~ height of chair
mile (m) ~ 15 min walk

Conversions

Larger to smaller units --- multiply
Smaller to larger units --- divide
1 kg = 1000 g
1 g = 1000 mg

kg → g   multiply by 1000
g → mg   multiply by 1000

mg → g   divide by 1000
g → kg   divide by 1000

km → m   multiply by 1000
m → cm   multiply by 100
cm → mm multiply by 10

mm → cm divide by 10
cm → m   divide by 100
m → km   divide by 10

Rules of Divisibility

Rules of divisibility by 2, 3, 5, 9 and 10

Factor	Test for divisibility
2	The ones digit is an even number.
3	The sum of the digits is divisible by 3.
5	The ones digit is a 0 or 5.
9	The sum of the digits is divisible by 9.
10	The ones digit is a 0.

Square Units

in^2 → ft^2 divide by 144
ft^2 → in^2 multiply by 144
ft^2 → yd^2 divide by 9
yd^2 → ft^2 multiply by 9

Other Helpful Measurements

16 oz = 1 lb
1 T = 2000 lb   (T= ton)
2 cups = 1 pint
4 quarts = 1 gallon
4 cups = 1 quart
8 pints = 1 gallon
2 pints = 1 quart
12 in = 1 ft
3 ft = 1 yd
5280 yd = 1 mile

Classifying Angles

Corresponding

Angles that are in the same position and are formed by a transversal cutting two or more parallel.

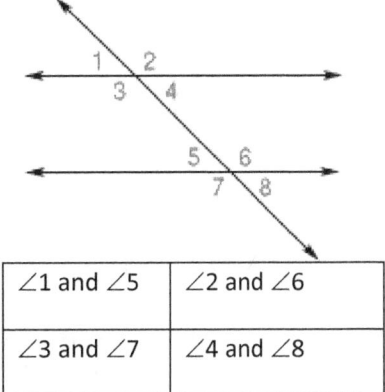

∠1 and ∠5	∠2 and ∠6
∠3 and ∠7	∠4 and ∠8

Vertical

A pair of opposite congruent angles formed by intersecting lines.

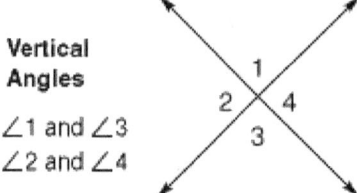

Vertical Angles

∠1 and ∠3
∠2 and ∠4

Adjacent

Angles that share a common side, have the same vertex, and do not overlap.

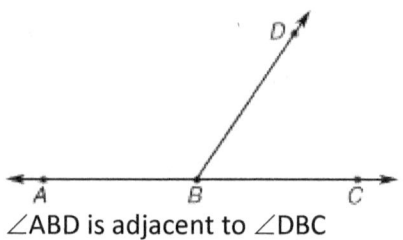

∠ABD is adjacent to ∠DBC

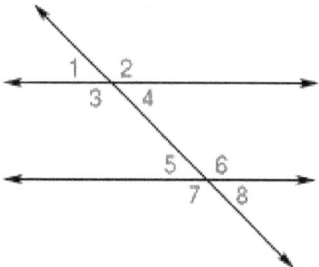

∠1 and ∠4	∠2 and ∠3
∠5 and ∠8	∠6 and ∠7

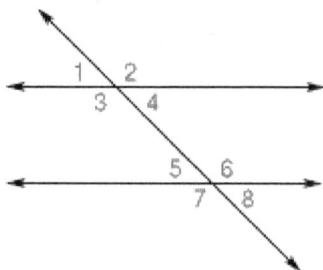

∠1 and ∠2	∠1 and ∠3
∠2 and ∠4	∠3 and ∠4
∠5 and ∠6	∠5 and ∠7
∠6 and ∠8	∠7 and ∠8

Complimentary Angles

Two angles whose measures have a sum of 90°

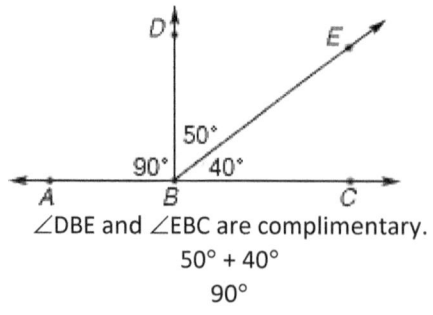

∠DBE and ∠EBC are complimentary.
50° + 40°
90°

Supplementary Angles

Two angles whose sum equals 180°.

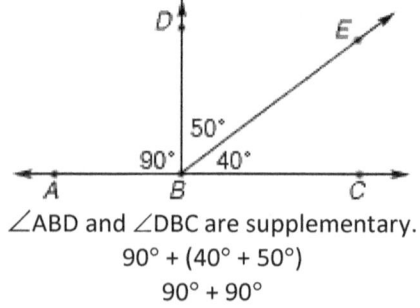

∠ABD and ∠DBC are supplementary.
90° + (40° + 50°)
90° + 90°
180°

Geometry Terminology

Angle A geometric figure formed by two rays that have a common endpoint.

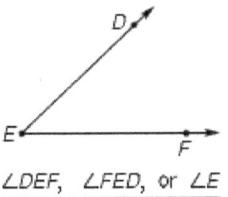

∠DEF, ∠FED, or ∠E

Congruent having the same size and shape.

Intersecting Lines lines that cross at exactly one poin

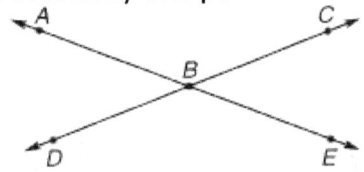

Line AE intersects line CD at point B.

Line A set of points that extends without end in opposite directions.

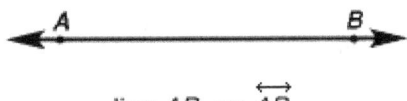

line AB, or $\overleftrightarrow{AB}$

Solid Figure a three-dimensional figure

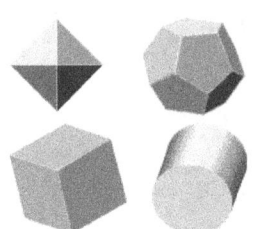

Transversal a line that intersects two or more lines

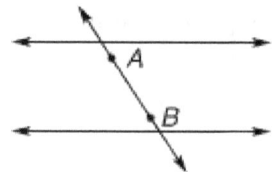

Line *AB* is a transversal.

Vertex The point where two or more rays meet; the point of intersection of two sides of a polygon; the point of intersection of three or more edges of a solid figure; the top point of a cone.

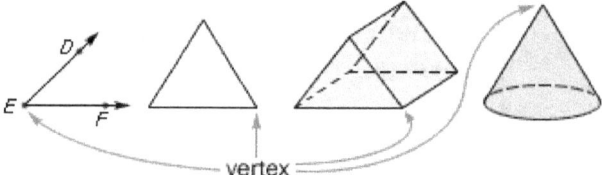

Ray A part of a line that has one endpoint and goes on forever in only one direction

Polygon A closed plane figure formed by three or more line segments

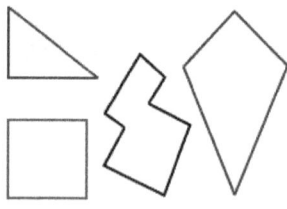

Plane Figure A figure which lies in a plane

Plane	A flat surface that goes on forever in all directions

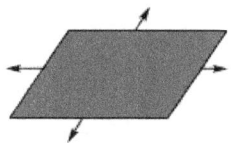

Edge	The line segment where two faces of a solid figure meet

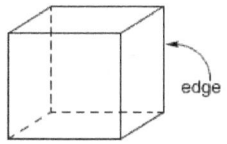

Line Segment	Part of a line with two endpoints

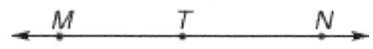

line segments *MT*, *TN*, and *MN*.
$\overline{MT}$, $\overline{TN}$, and $\overline{MN}$

Face	One of the polygons of a solid figure

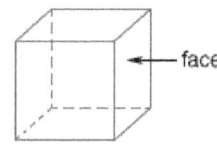

The cube has 6 faces.

Three-Dimensional	Having length, width, and height

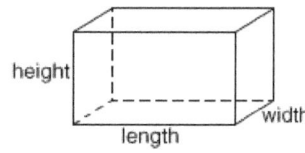

The rectangular prism is three-dimensional.

Cone   A solid figure with a circular base and one vertex

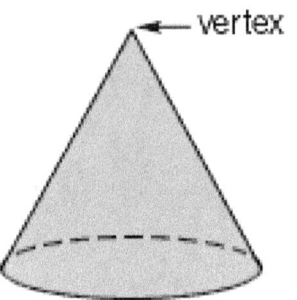

Base   A side of a polygon or a face of a solid figure by which the figure is measured or named

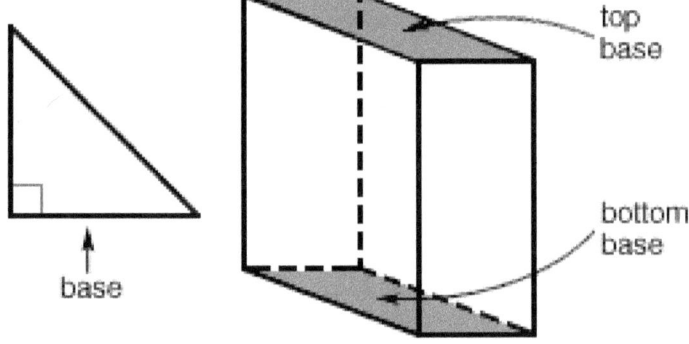

PARCC Practice Test 1

Partnership for Assessment of Readiness for College and Careers

Grade 7

Mathematics

2018

Calculators are NOT permitted for unit 1 of the test.

Time for Unit 1: 80 Minutes

Unit 1

1) In a party, 10 soft drinks are required for every 12 guests. If there are 252 guests, how many soft drink is required?

   A. 21
   B. 105
   C. 210
   D. 2510

2) You can buy 5 cans of green beans at a supermarket for $3.40. How much does it cost to buy 35 cans of green beans?

   A. $17
   B. $23.80
   C. $34.00
   D. $119

3) 1.2 is what percent of 24?

   A. 1.2
   B. 5
   C. 12
   D. 24

4) Joe scored 20 out of 25 marks in Algebra, 30 out of 40 marks in science and 68 out of 80 marks in mathematics. In which subject his percentage of marks is best?

   A. Algebra
   B. Science
   C. Mathematics
   D. Algebra and Science

5) What is the volume of the following square pyramid?

   A. 120 m³
   B. 144 m³
   C. 480 m³
   D. 1440 m³

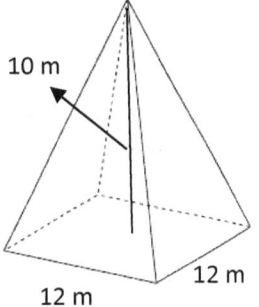

6) What is the area of the shaded region?

   A. 9 π cm2
   B. 25 π cm2
   C. 39 π cm2
   D. 64 π cm2

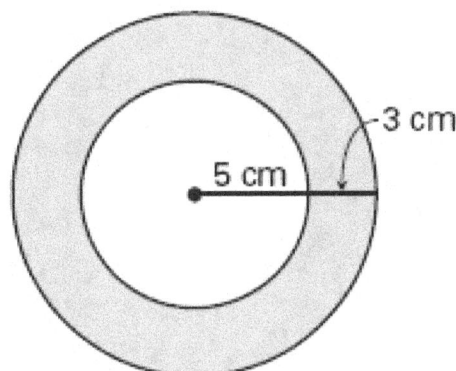

7) A pizza cut into 8 parts. William and his sister Ella ordered two pizzas. William ate $\frac{1}{4}$ of his pizza and Ella ate $\frac{1}{2}$ of her pizza. What part of the two pizzas was left?

   A. $\frac{1}{2}$

   B. $\frac{1}{3}$

   C. $\frac{3}{8}$

   D. $\frac{5}{8}$

8) Robert is preparing to run a marathon. He runs $7\frac{1}{5}$ miles on Saturday and two times that many on Monday and Wednesday. Robert wants to run a total of 60 miles this week. How many more miles does he need to run?
   Write your answer in the box below.

9) Simplify $6x^2y^3(2x^2y)^3 =$

   A. $12x^4y^6$

   B. $12x^8y^6$

   C. $48x^4y^6$

   D. $48x^8y^6$

10) 5 less than twice a positive integer is 83. What is the integer?

   A. 39
   B. 41
   C. 42
   D. 44

11) Which of the following points lies on the line $4x + 6y = 14$?

   A. (2, 1)
   B. (-1, 3)
   C. (-3, 4)
   D. (2, 2)

12) How many possible outfit combinations come from six shirts, three slacks, and five ties?

   Write your answer in the box below.

   ☐

13) An angle is equal to one fifth of its supplement. What is the measure of that angle?

   A. 20
   B. 30
   C. 45
   D. 60

14) John traveled 150 km in 6 hours and Alice traveled 180 km in 4 hours. What is the ratio of the average speed of John to average speed of Alice?
   A. 3 : 2
   B. 2 : 3
   C. 5 : 9
   D. 5 : 6

15) If 40 % of a class are girls, and 25 % of girls play tennis, what fraction of the class play tennis?
   A. 10 %
   B. 15%
   C. 20 %
   D. 40 %

16) Right triangle ABC has two legs of lengths 6 cm (AB) and 8 cm (AC). What is the length of the third side (BC)?
   A. 4 cm
   B. 6 cm
   C. 8 cm
   D. 10 cm

17) The marked price of a computer is D dollar. Its price decreased by 20% in January and later increased by 10 % in February. What is the final price of the computer in D dollar?
   A. 0.80 D
   B. 0.88 D
   C. 0.90 D
   D. 1.20 D

This is the end of Unit 1

Scientific Calculators are permitted for unit 2 of the practice test.

Time for Unit 2: 80 Minutes

Unit 2

18) [6 × (-24) + 8] − (-4) + [4 × 5] ÷ 2 = ?

Write your answer in the box below.

☐

19) The area of a circle is 64 π. What is the circumference of the circle?
   A. 8 π
   B. 16 π
   C. 32 π
   D. 64 π

20) A $40 shirt now selling for $28 is discounted by what percent?
   A. 20 %
   B. 30 %
   C. 40 %
   D. 60 %

21) From last year, the price of gasoline has increased from $1.25 per gallon to $1.75 per gallon. The new price is what percent of the original price?
   A. 72 %
   B. 120 %
   C. 140 %
   D. 160 %

22) A boat sails 40 miles south and then 30 miles east. How far is the boat from its start point?
   A. 45
   B. 50
   C. 60
   D. 70

23) Sophia purchased a sofa for $530.40. The sofa is regularly priced at $624. What was the percent discount Sophia received on the sofa?
   A. 12%
   B. 15%
   C. 20%
   D. 25%

24) The score of Emma was half as that of Ava and the score of Mia was twice that of Ava. If the score of Mia was 60, what is the score of Emma?
   A. 12
   B. 15
   C. 20
   D. 30

25) A bag contains 18 balls: two green, five black, eight blue, a brown, a red and one white. If 17 balls are removed from the bag at random, what is the probability that a brown ball has been removed?

A. $\frac{1}{9}$

B. $\frac{1}{6}$

C. $\frac{16}{17}$

D. $\frac{17}{18}$

26) A rope weighs 600 grams per meter of length. What is the weight in kilograms of 12.2 meters of this rope? (1 kilograms = 1000 grams)

A. 0.0732

B. 0.732

C. 7.32

D. 7,320

27) A chemical solution contains 4% alcohol. If there is 24 ml of alcohol, what is the volume of the solution?

A. 240 ml

B. 480 ml

C. 600 ml

D. 1200 ml

28) The price of a laptop is decreased by 10% to $360. What is its original price?

A. 320

B. 380

C. 400

D. 450

29) What is the median of these numbers? 4, 9, 13, 8, 15, 18, 5

A. 8

B. 9

C. 13

D. 15

30) Three times the price of a laptop is equal to five times the price of a computer. If the price of laptop is $200 more than the computer, what is the price of the computer?

A. 300

B. 500

C. 800

D. 1500

This is the end of Unit 2

Scientific Calculators are permitted for unit 3 of the practice test.

Time for Unit 2: 80 Minutes

Unit 3

31) What is the perimeter of a square that has an area of 595.36 feet?

Write your answer in the box below.

☐

32) Jason is 9 miles ahead of Joe running at 5.5 miles per hour and Joe is running at the speed of 7 miles per hour. How long does it take Joe to catch Jason?
   A. 3 hours
   B. 4 hours
   C. 6 hours
   D. 8 hours

33) 55 students took an exam and 11 of them failed. What percent of the students passed the exam?
   A. 20 %
   B. 40 %
   C. 60 %
   D. 80 %

34) Jason needs an 75% average in his writing class to pass. On his first 4 exams, he earned scores of 68%, 72%, 85%, and 90%. What is the minimum score Jason can earn on his fifth and final test to pass?

Write your answer in the box below.

☐

35) A bank is offering 3.5% simple interest on a savings account. If you deposit $12,000, how much interest will you earn in two years?
   A. $420
   B. $840
   C. $4200
   D. $8400

36) What is the median of these numbers? 2, 28, 28, 19, 67, 44, 35
   A. 19
   B. 28
   C. 44
   D. 35

37) Last week 24,000 fans attended a football match. This week three times as many bought tickets, but one sixth of them cancelled their tickets. How many are attending this week?
   A. 48,000
   B. 54,000
   C. 60,000
   D. 72,000

38) The following trapezoid are similar. What is the value of x ?

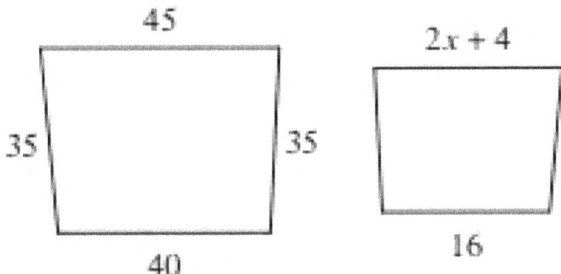

- A. 7
- B. 8
- C. 18
- D. 45

39) If $x = -8$, which equation is true?

- A. $x(2x - 4) = 120$
- B. $8(4 - x) = 96$
- C. $2(4x + 6) = 79$
- D. $6x - 2 = -46$

40) In a bag of small balls $\frac{1}{3}$ are black, $\frac{1}{6}$ are white, $\frac{1}{4}$ are red and the remaining 12 blue. How many balls are white?
- A. 8
- B. 12
- C. 16
- D. 24

This is the end of Unit 3

PARCC Practice Test 2

Partnership for Assessment of Readiness for College and Careers

Grade 7

Mathematics

2018

Calculators are NOT permitted for unit 1 of the test.

Time for Unit 1: 80 Minutes

Unit 1

1) 11 yards 6 feet and 4 inches equals to how many inches?

   A. 388
   B. 468
   C. 472
   D. 476

2) A shirt costing $200 is discounted 15%. After a month, the shirt is discounted another 15%. Which of the following expressions can be used to find the selling price of the shirt?

   A. (200) (0.70)
   B. (200) − 200 (0.30)
   C. (200) (0.15) − (200) (0.15)
   D. (200) (0.85) (0.85)

3) Which of the following points lies on the line 2x + 4y = 10
   A. (2, 1)
   B. (-1, 3)
   C. (-2, 2)
   D. (2, 2)

4) 5 + 8 × (-2) − [4 + 22 ×5] ÷ 6 = ?

Write your answer in the box below.

5) The price of a car was $20,000 in 2014, $16,000 in 2015 and $12,800 in 2016. What is the rate of depreciation of the price of car per year?

   A. 15 %
   B. 20 %
   C. 25 %
   D. 30 %

6) The width of a box is one third of its length. The height of the box is one third of its width. If the length of the box is 27 cm, what is the volume of the box?

   A. 81 cm3
   B. 162 cm3
   C. 243 cm3
   D. 729 cm3

7) In five successive hours, a car travels 40 km, 45 km, 50 km, 35 km and 55 km. In the next five hours, it travels with an average speed of 50 km per hour. Find the total distance the car traveled in 10 hours.

   A. 425 km
   B. 450 km
   C. 475 km
   D. 500 km

8) The ratio of boys to girls in a school is 2:3. If there are 600 students in a school, how many boys are in the school.

Write your answer in the box below.

9) The perimeter of the trapezoid below is 54 cm. What is its area?

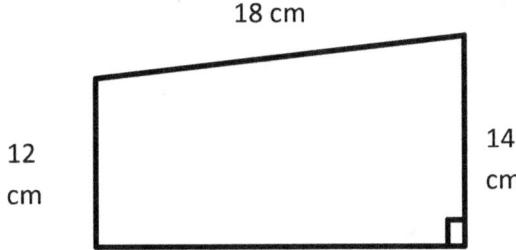

Write your answer in the box below.

10) In 1999, the average worker's income increased $2,000 per year starting from $24,000 annual salary. Which equation represents income greater than average? (I = income, x = number of years after 1999)

   A. I > 2000 x + 24000

   B. I > - 2000 x + 24000

   C. I < -2000 x + 24000

   D. I < 2000 x - 24000

11) Which of the following graphs represents the compound inequality $-2 \leq 2x-4 < 8$?

   A.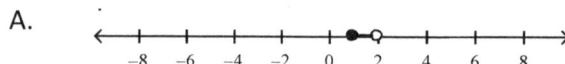

   B.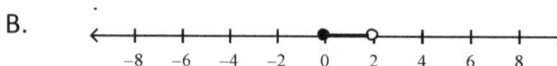

   C.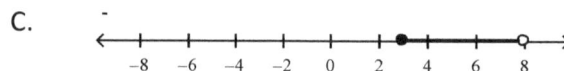

   D.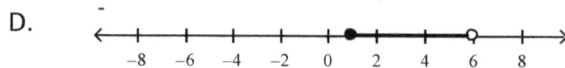

12) A football team had $20,000 to spend on supplies. The team spent $14,000 on new balls. New sport shoes cost $120 each. Which of the following inequalities represent how many new shoes the team can purchase.

   A. $120x + 14,000 \leq 20,000$
   B. $120x + 14,000 \geq 20,000$
   C. $14,000x + 12,0 \leq 20,000$
   D. $14,000x + 12,0 \geq 20,000$

13) Two dice are thrown simultaneously, what is the probability of getting a sum of 6 or 9?

   A. $\dfrac{1}{3}$
   B. $\dfrac{1}{4}$
   C. $\dfrac{1}{6}$
   D. $\dfrac{1}{12}$

14) A swimming pool holds 2,000 cubic feet of water. The swimming pool is 25 feet long and 10 feet wide. How deep is the swimming pool?

   Write your answer in the box below.

15) When a number is subtracted from 24 and the difference is divided by that number, the result is 3. What is the value of the number?

   A. 2
   B. 4
   C. 6
   D. 12

16) What is the volume of a box with the following dimensions?

Hight = 4 cm   Width = 5 cm   Length = 6 cm

   A. 15 cm3
   B. 60 cm3
   C. 90 cm3
   D. 120 cm3

17) In two successive years, the population of a town is increased by 15% and 20%. What percent of its population is increased after two years?

   A. 32
   B. 35
   C. 38
   D. 68

18) In a school, the ratio of number of boys to girls is 4:5. If the number of boys is 180, what is the total number of students in the school? Write your answer in the box below.

19) How many tiles of 8 cm^2 is needed to cover a floor of dimension 6 cm by 24 cm?

   A. 6
   B. 12
   C. 18
   D. 24

20) The radius of the following cylinder is 8 inches and its height is 12 inches. What is the surface area of the cylinder?

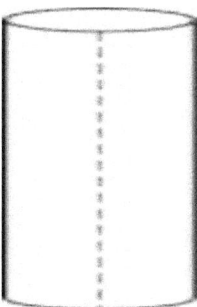

   A. 96 π cm^2
   B. 192 π cm^2
   C. 320 π cm^2
   D. 768 π cm^2

This is the end of Unit 1

Scientific Calculators are permitted for unit 2 of the practice test.

Time for Unit 2: 80 Minutes

Unit 2

21) Which graph corresponds to the following inequalities?

$y \leq x + 4$

$2x + y \leq -4$

A.   B.

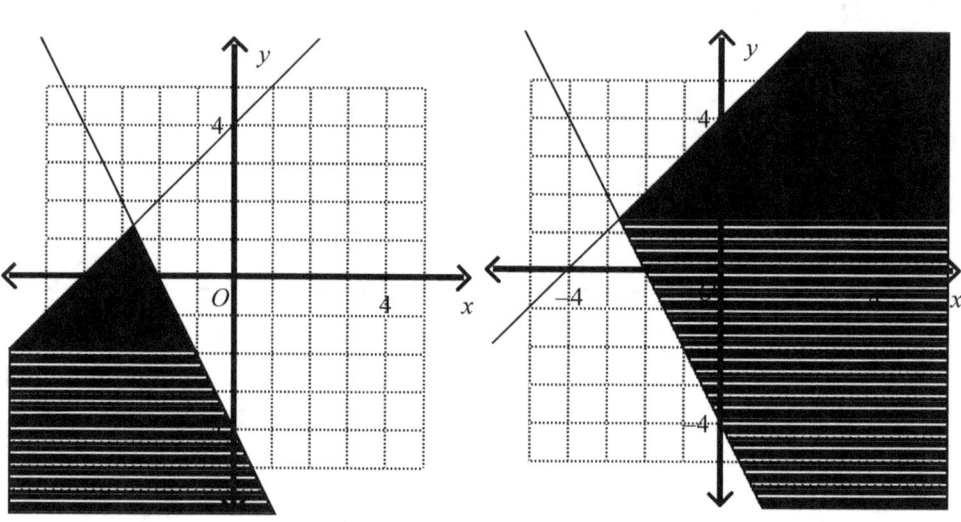

C.   D.

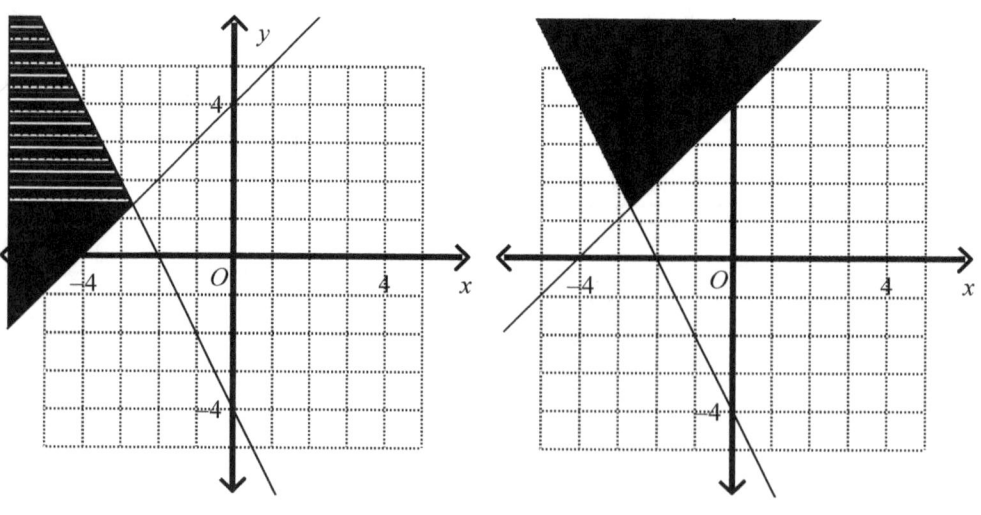

22) A bank is offering 4.5% simple interest on a savings account. If you deposit $8,000, how much interest will you earn in five years?

    A. $360

    B. $720

    C. $1800

    D. $3600

23) A card is drawn at random from a standard 52-card deck, what is the probability that the card is of Hearts? (The deck includes 13 of each suit clubs, diamonds, hearts, and spades)

    A. $\frac{1}{3}$

    B. $\frac{1}{4}$

    C. $\frac{1}{6}$

    D. $\frac{1}{52}$

24) How long does a 420-miles trip take moving at 50 miles per hour (mph)?

   A. 4 hours
   B. 6 hours and 24 minutes
   C. 8 hours and 24 minutes
   D. 8 hours and 30 minutes

25) What is the equivalent temperature of 104°F in Celsius?
$$C = \frac{5}{9}(F - 32)$$

   A. 32
   B. 40
   C. 48
   D. 68

26) The square of a number is $\frac{25}{64}$. What is the cube of that number?

   A. $\frac{5}{8}$
   B. $\frac{25}{254}$
   C. $\frac{125}{512}$
   D. $\frac{125}{64}$

27) What is the surface area of the cylinder below?

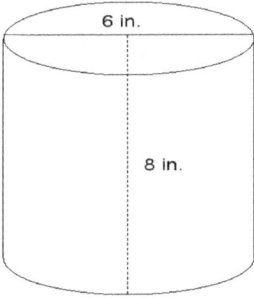

A. 48π

B. 57π

C. 66π

D. 288π

28) What is the value of x in the following equation?

$$\frac{2}{3}x + \frac{1}{6} = \frac{1}{3}$$

A. 6

B. $\frac{1}{2}$

C. $\frac{1}{3}$

D. $\frac{1}{4}$

29) The average of five numbers is 24. If a sixth number 42 is added, then, what is the new average?

A. 25

B. 26

C. 27

D. 28

30) Anita's trick-or-treat bag contains 12 pieces of chocolate, 18 suckers, 18 pieces of gum, 24 pieces of licorice. If she randomly pulls a piece of candy from her bag, what is the probability of her pulling out a piece of sucker?

A. $\frac{1}{3}$

B. $\frac{1}{4}$

C. $\frac{1}{6}$

D. $\frac{1}{12}$

This is the end of Unit 2

Scientific Calculators are permitted for unit 3 of the practice test.

Time for Unit 3: 80 Minutes

Unit 3

31) Which of the following shows the numbers in descending order?

$\frac{2}{3}, 0.68, 67\%, \frac{4}{5}$

A. $67\%, 0.68, \frac{2}{3}, \frac{4}{5}$

B. $67\%, 0.68, \frac{4}{5}, \frac{2}{3}$

C. $0.68, 67\%, \frac{2}{3}, \frac{4}{5}$

D. $\frac{2}{3}, 67\%, 0.68, \frac{4}{5}$

32) What is the slope of a line that is perpendicular to the line $4x - 2y = 12$?

A. -2
B. 2
C. 4
D. 12

33) What is the value of the expression $5(x - 2y) + (2 - x)^2$ when $x = 3$ and $y = -2$?

A. -4
B. 20
C. 36
D. 50

34) The mean of 50 test scores was calculated as 88. But, it turned out that one of the scores was misread as 94 but it was 69. What is the mean?

A. 85

B. 87

C. 87.5

D. 88.5

35) Mr. Carlos family are choosing a menu for their reception. They have 3 choices of appetizers, 5 choices of entrees, 4 choices of cake. How many different menu combinations are possible for them to choose?

   A. 12

   B. 32

   C. 60

   D. 120

36) Four one - foot rulers can be split among how many users to leave each with $\frac{1}{6}$ of a ruler?

   A. 4

   B. 6

   C. 12

   D. 24

37) What is the area of a square whose diagonal is 8?

   A. 16

   B. 32

   C. 36

   D. 64

38) The ratio of boys and girls in a class is 4:7. If there are 44 students in the class, how many more boys should be enrolled to make the ratio 1:1?

A. 8

B. 10

C. 12

D. 14

39) What is the area of the shaded region?

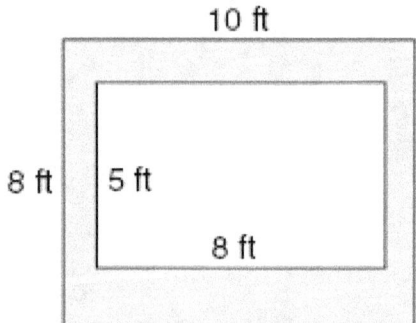

A. 31

B. 40

C. 64

D. 80

40) Mr. Jones saves $2,500 out of his monthly family income of $55,000. What fractional part of his income does he save?

A. $\frac{1}{22}$

B. $\frac{1}{12}$

C. $\frac{3}{25}$

D. $\frac{2}{15}$

This is the end of Unit 3

PARCC Practice Tests Answer keys

\multicolumn{4}{c	}{PARCC Practice Test 1 Answer Key}	\multicolumn{4}{c}{PARCC Practice Test 2 Answer Key}
1.	C	2.	B	1.	C	2.	D
3.	B	4.	C	3.	B	4.	-30
5.	C	6.	C	5.	B	6.	D
7.	D	8.	24	7.	C	8.	240
9.	D	10.	D	9.	130	10.	A
11.	B	12.	90	11.	D	12.	A
13.	B	14.	C	13.	D	14.	8
15.	A	16.	D	15.	C	16.	D
17.	B	18.	-122	17.	C	18.	405
19.	B	20.	B	19.	C	20.	C
21.	C	22.	B	21.	A	22.	C
23.	B	24.	B	23.	B	24.	C
25.	D	26.	C	25.	B	26.	C
27.	C	28.	C	27.	C	28.	D
29.	B	30.	A	29.	C	30.	B
31.	97.6	32.	C	31.	D	32.	B
33.	D	34.	60	33.	C	34.	C
35.	B	36.	B	35.	C	36.	D
37.	C	38.	A	37.	B	38.	C
39.	B	40.	A	39.	B	40.	A

"Effortless Math Education" Publications

Effortless Math Education authors' team strives to prepare and publish the best quality Mathematics learning resources to make learning Math easier for all. We hope that our publications help you or your student learn Math in an effective way.

We all in Effortless Math wish you good luck and successful studies!

Effortless Math Authors

Online Math Lesso

Enjoy interactive Math lessons online with the best Math teachers

Online Math learning that's e

affordable, flexible, and

Learn Math wherever you want; when you want

Ultimate flexibility. You can now learn Math online, enjoy high quality engaging lessons no matter where in the world you are. It's affordable too.

Learn Math with one-on-one classes

We provide one-on-one Math tutoring online. We believe that one-to-one tutoring is the mos effective way to lea Math.

Online Math L

It's easy! Here's ho

1- Request a FREE intro

2- Meet a Math tutor o

3- Start Learning Mat

Send Email to: info@Effortles

Or Call: +1-469-230-36